Volume 14

Ceramic Transactions

ELECTRO-OPTICS AND NONLINEAR OPTIC MATERIALS

Edited by A.S. Bhalla, Pennsylvania State University,
E.M. Vogel, Bell Communications Research, and
K.M. Nair, E.I. du Pont de Nemours & Company, Inc.

The American Ceramic Society, Inc.
Westerville, Ohio

Proceedings of the Symposium on Electro-Optics and Nonlinear Optics held in Anaheim, California, October 31–November 3, 1989.

Library of Congress
Cataloging-in-Publication Data

Symposium on Electro-Optics and Nonlinear Optics (1989 : Anaheim, Calif.)
Electro-optics and nonlinear optic materials.
(Ceramic transactions ; v. 14)
"Proceedings of the Symposium on Electro-Optics and Nonlinear Optics, held in Anaheim, California, October 31–November 3, 1989" — T.p. Verso.
Includes index.
1. Electrooptic—Congresses. 2. Electrooptical devices—Congresses.
I. Bhalla, A. A., 1943- . II. Vogel, E. M. III. Nair, K. M.
IV. American Ceramic Society. V. Title. VI. Series.
TA1750.S963 1989 621.3815 90-1227
ISBN 0-944904-32-7

Printed in the United States of America

1 2 3 4—93 92 91 90

Preface

The activity in the area of electronics, particularly microelectronics, has been one of the major scientific and technical frontiers for the last three decades. The complex technology development in this area emerged and has been appreciated only a decade later, after the invention of the transistor in 1948 and integrated circuits (ICs) soon after. Various materials and processing technologies were developed in the area of electronics during this golden age, and they have made great strides in almost all facets: consumer electronics, telecommunications, medical electronics, and electronics for automotive, industrial production, and military applications. However, it is a fast changing technology and new materials and processes are needed for improved device applications. Many scientists and engineers are in the forefront of this area, developing innovative processing and materials technologies.

The decision of the American Ceramic Society to organize several international symposia covering many aspects of advanced electronic material systems by bringing together leading researchers and practitioners of electronics and to publish the proceedings of the conferences in the Ceramic Transactions series was a major contribution to the electronics industry. As a result, symposia on Materials and Processes for Microelectronic Systems, Electro-Optics and Nonlinear Optic Materials, Glasses for Electronic Applications, and Ceramic Superconductor Materials and Processes were organized during the first and second Ceramic Science and Technology Congresses, sponsored by the Society.

This volume contains a collection of papers presented at the international symposium on Electro-Optics and Nonlinear Optic Materials, during the First International Science and Technology Congress, held at Anaheim, CA, October 31–November 3, 1989. The major topics of the symposium were Nonlinear Optical Materials and Devices; Electro-Optical, Acousto-Optical, Magneto-Optical Devices and Applications; Photosensitive Materials and Devices; Electro-Optical Instruments and Devices; and Optical Fibers and Optical Disks. Over seventeen invited papers and many contributed papers were presented during this three-day conference. All the papers from the conference that are included in this volume were reviewed prior to publication by experts in the area.

We, the editors, acknowledge and appreciate the contributions of the speakers, conference session chairs, manuscript reviewers, and society officials for making this endeavor a successful one.

A.S. Bhalla
E.M. Vogel
K.M. Nair

Ceramic Transactions *is a new proceedings series designed to meet two needs: high quality content and rapid publication. Volumes in the series come from meetings, symposia, and forums. Each paper is reviewed by two peers, and final manuscripts are prepared by authors in a ''camera-ready'' format. The volumes in this series would not be possible without the hard work, dedication, and cooperation of editors, reviewers, and authors, who all deserve a great deal of thanks.*

Your comments, questions, and suggestions for future ***Ceramic Transactions*** *volumes are welcomed and should be addressed to the Director of Publications, The American Ceramic Society, Inc.*

Contents

Section I. Electro-Optic Materials

Section II. Second and Third Order Nonlinear Optical Effects

Section III. Optical Thin Films and Applications

Section IV. Device Applications

Section I. Electro-Optic Materials

FERROELECTRIC ELECTRO-OPTIC MATERIALS

Frank W Ainger
Plessey Research Caswell Ltd
Caswell, Towcester
Northants NN12 8EQ, UK

ABSTRACT

The need for non-linear optical materials has revived interest in ferroelectric oxides, a number of which have been revisited to establish their suitability and applicability. Many are multi-component systems and pose a significant challenge to crystal growers. However, some of these problems have been resolved by considering alternatives, such as optical transparent ceramics and thin films. Current trends will be reviewed and future prospects postulated.

INTRODUCTION

The demand for electro-optic materials has significantly grown in the past decade, having been driven by the rapid advances in fiber optic systems and opto-electronics for telecommunications signal processing, sensors and modulators as indicated in Table 1.

Table 1. Applications for Electro-Optic Materials

Modulator Arrays	Fiber Optics
Spatial Light Modulators Parallel Processing Optical Computing Displays Recording Media	Signal Processing Telecommunications Sensors Radar

Ferroelectrics containing the BO_6 oxygen octahedra often exhibit high dielectric constants, refractive indices and spontaneous polarisations and have therefore been targetted for further development as electro-optic materials and the subject of this review.

Di Domenico and Wemple[1,2] presented a unified theory for the E O and NLO effects in the class of oxygen octahedral ferro-electrics based on an understanding of their electronic energy band structure. It emphasises the importance of the basic BO_6 building block which is found in simple and distorted perovskites, tungsten bronze and related structures, bismuth oxide layer compounds and lithium niobate structure types.

The BO_6 oxygen octahedral governs the low lying conduction bands and the upper valance bands and its significance is that we expect them to be similar in all oxygen octahedral ferroelectrics since the B cation d-orbitals and oxygen anion 2p orbitals are the major contributors to the energy bands of interest. Whilst other ions may contribute to the higher lying conduction bands, they are considered to have negligible importance provided their electronic polarisability is small. The exceptions are Bi^{3+} and Pb^{2+}. The polarisation potential has been determined to be the same in all oxygen octahedral ferroelectrics and it can be concluded that the clamped quadratic electro-optic g-coefficients associated with each BO_6 unit are essentially identical and the EO and NLO tensor coefficients are fundamentally the same.

Further details of how the relevant relationships are derived and used fall outside the scope of this paper. Results shown in table 2 show how BO_6 ferroelectrics qualify for some of the applications and modes of operation, but it is first worth showing how the non-linear coefficients are related to the polar properties.

NON-LINEAR OPTICS

Non-linear optics is an extensive topic so a fairly narrow view has been taken in a review paper on materials. Two of the most important aspects are optical frequency mixing and the electro-optic effect. In the former, we are concerned with the generation of sum and difference optical frequencies (when radiation falls on a crystal) whereas the electro-optic effect is the change of refractive index in a material induced by an applied field. Such materials are said to be non-linear and the field may originate from the light source itself or be applied.

TABLE 2

Dielectric properties and Electro-optic Properties of BO_6 Ferroelectrics at Room Temp. & 0.633μm.

Material	Tc ^{0}C	ε_c	P_s C.m-2	εP_s	V_π(volts)
$LiTaO_3$	610	45	0.71	22	2500
$LiNbO_3$	1200	30	0.50	21	2600
$Ba_4Na_2Nb_{10}O_{30}$	560	51	0.40	20	1600
$Sr_4K_2Nb_{10}O_{30}$	156	400	0.25	100	400
$K_6Li_4Nb_{10}O_{30}$	420	100	0.25	25	930
$Sr_{2.5}Ba_{2.5}Nb_{10}O_{30}$	115	300	0.25	75	400
$BaTiO_3$	134	150	0.25	38	500
$KTa_{0.65}Nb_{0.35}O$	15	8000	0.05	400	50

For a linear electro-optic material, the induced polarisation, P, is directly proportional to the field E.

$$P = \varepsilon_0 \chi E$$

where ε_0 = permittivity of free space
χ = susceptibility of the materials and is related to the refractive index at optical wavelength by

$$n^2 = 1 + \chi$$

The non-linear relationship and polarisation is related to E by a series expansion.

$$P = \varepsilon_0 (\chi_{(1)}E + \chi_{(2)}E^2 + \chi_{(3)}E^3 + \ldots)$$

For a centrosymmetric material $\chi_{(2)} = 0$ and therefore to retain this imporant non-linear term, we require crystals which are acentric, ie, lack a centre of symmetry.

If we introduce, in a simplified form, an oscillating wave for the lightwave:

$$E = E_0 \sin 2\pi ft$$

For a linear material

$$P = \varepsilon_0 \chi E_0 \sin 2\pi ft = P_0 \sin 2\pi ft$$

and P oscillates at the same frequency.

In the non-linear case

$$P = \varepsilon_0\chi_{(1)}E_0 \sin 2\pi ft + \varepsilon_0\chi^2[E_0 \mathrm{Sin} 2\pi ft]^2 + \ldots$$

or $$P = \varepsilon_0\chi_{(1)}E_0 \sin 2\pi ft + \varepsilon_0\chi^2 E_0^2\frac{[1 - \mathrm{Cos}\ 4\pi ft]}{2}$$

showing a component oscillating a frequency of 2f. This is the special case known as second harmonic generation (SHG).

The second order effects are important to all of the current applications. The first of these E-O effects is the Pockels Effect, where an applied field changes the refracive index of the medium which in turn modulates the optical wave. This is the basis of fast electro-optic shutters (Q-switches) and there is a demand for more non-linear materials operating at faster speeds and lower voltages. Similar materials are also needed for more efficient optical waveguides for rerouting optical signals into optical fibers.

Optical frequency mixing is important and gives rise through the $\chi_{(2)}E^2$ term to polarisation waves of which that light incident at f and emerging from the material at 2f is known as second harmonic generation. It is valuable for doubling the frequency of solid state lasers operating at a fundamental frequency in the near infra-red so that green light is obtained. Its implications for the consumers can be easily recognised in the case of the compact disc. Four times as much information could be stored and read with the possibility of 4 hours of digital sound stored in a 10cm disc.

When $\chi_{(2)}E^2$ is zero as in the case of centrosymmetric materials, the quadratic electro-optic effect known as the Kerr Effect is derived from the $\chi_{(3)}E^3$ term. In this case, the refractive index and phase changes are related to the squares of the applied fields.

Kerr cells have been effectively constructed from the lead lanthanum zirconium titanate (PLZT) electro-optic ceramics which are pseudocubic and show a strong effect under an applied field.

In the case of optical frequency mixing and SHG, a large value of $\chi_{(2)}$ is required, which is empirically related to the refractive index by $\chi_{(2)} \alpha (n^2 - 1)^3$, suggesting that large values of n are desirable. In the electro-optic case large non-linearity again demands high values of $\chi_{(2)}$ but to achieve low working voltages compounds with high RF dielectric constants are preferred.

REQUIREMENTS FOR ELECTRO-OPTIC MATERIAL

Firstly, we have to establish guidelines as to the properties of the crystals required. The most important properties are cited below, but do not include any special features needed for a specific application. However, if these conditions are fulfilled by our chosen material, then we are likely to succeed in its specified application. Our ferroelectric crystal or material needs to be:

1. non-centrosymmetric otherwise significant non-linear effects disappear.

2. transparent in that part of the spectrum of interest usually 0.4-5μm.

3. chemically homogeneous which by definition generally excludes solid solution and mixed crystals (for electro-optic ceramics solid solution effects can be exploited; this special case will be dealt with separately).

4. readily grown by an available technique to the required size and optical quality.

5. grown as, or subsequently poled to, a stable single domain state.

6. cut and polished to interferometric optical quality.

7. durable in normal ambient conditions and not suffer damage when exposed to the optical beam.

Although there exist points of difference for SHG and electro-optic effects (the former requiring optical anisotropy to provide phase matching whilst the latter prefers optical isotropy to avoid temperature control), there is an underlying unity for both aspects of non-linear optics.

From these requirements, there is no direct link to the kind of chemical compositions which would fully satisfy them. However, there are clues eluded to earlier which suggest than non-centro-symmetric ferroelectrics with high dielectric constants in the Curie region and useful electro-optic coefficients derived from the Di Domenico and Wemple [1,2] analysis may qualify. The most common oxygen octahdedral ferroelectrics are the niobates, tantalates, titanates and zirconates which are chemically stable and will survive the environmental conditions imposed by the application. However, many of them are difficult to grow, so even if we were able to predict their chemical constitution, their quality as grown crystals would not be known.

Most of the oxides referred to above have wide band gaps with corresponding optical transmissions in the spectral range of 0.4-5μm, which fulfil many of the current device applications.

As will be seen later, the growth of ferroelectric oxide single crystals poses some considerable technological problems. This has challenged us to seek alternative forms to meet the need for some optical quality components and promoted research into bulk ceramics and thin films. The limited but significant success of these approaches will be described later suffice it to say it has afforded us another exploitable parameter, the solid solution effect, which is very limited in the case of single crystals.

FERROELECTRIC OXIDE MATERIALS

The choice of materials is bewilderingly large, but using our selection guidelines, it can be narrowed down to specific systems and examples contained within them. The current discussion will limit itself to four different structure types which have dominated materials research and optical device considerations for nearly twenty years.

Perovskites

The general formula of this family is ABO_3, of which barium titanate ($BaTiO_3$) is a pre-eminent example. The high temperature cubic phase is non-polar from which three polar pseudo-cubic distortions, tetragonal, rhombohedral and orthorhombic are derived. In some perovskite systems, there are two distinct rhombohedral phases related by a first order transition involving the tilting of the BO_6 octahedra. The polar directions found in these phases are: (a) six equivalent <100> axes in the tetragonal case, (b) eight equivalent <111> axes in the rhombohedral case, and (c) twelve equivalent <110> axes in the orthorhombic case. This polar behaviour can have a profound influence on the domain configurations and incipient twinning that may arise in crystals. Table 3 lists some of the materials which have been considered and studied for electo-optic applications, either as single crystals or transparent ceramics.

Table 3. Ferroelectrics for E-O Applications

Perovskites	(Solid Solution)
Barium Titanate $BaTiO_3$	$Ba_{1-x}Sr_xTiO_3$
Potassium Niobate $KNbO_3$	$KNb_{1-x}Ta_xO_3$
Lead Titanate $PbTiO_3$	$PbTi_{1-x}Zr_xO_3$
Lead Zirconate $PbZrO_3$	$Pb_{1-3/2}La_xTi_{1-y}Zr_yO_3$
Lead Magnesium Niobate $PbMg_{1/3}Nb_{2/3}O_3$	$(PbMg_{1/3}Nb_{2/3}O_3)_x\ (PbTiO_3)_{1-x}$
Cesium Germanium Chloride $CsGeCl_3$	$CsGe_{1-x}Pb_xCl_3$ $Cs_{1-x}Rb_xGeCl_3$

Single Crystal Perovskites

The growth of stoichiometric compounds, eg, $BaTiO_3$, $PbTiO_3$ and $KNbO_3$ has had some limited sucess by both melt and solution techniques and a study of $BaTiO_3$ will be made. The solid solution compounds $KTa_xNb_{1-x}O_3$ (KTN) and $Ba_xSr_{1-x}TiO_3$ (BST) with $x \sim 0.65$ have Curie temperature (T_C) near or at room temperature and attempts to grow large cubic crystals have failed to yield any substantial volumes of material suitable for optical devices. One of the major problems is that of chemical inhomogeneity. Variations of Tc as large as 100^0C/cm of crystal have been found in KTN. Even attempts at growth with 20kg melts/solutions in large platinum crucibles (buckets) have yielded crystals from which only small stress-free pieces suitable for research studies and limited characterisation are obtained.

Growth of $BaTiO_3$ From Solution

Barium titanate melts congruently at 1635^0C but the resulting hexagonal phase is a stable non-ferroelectric to room temperature. The transition to the cubic and desired phase is sluggish and therefore solution techniques which yield the cubic phase are required. Careful study of the $BaO-TiO_2$ phase diagram[3], Figure 1, is rewarding as it shows that cubic $BaTiO_3$ is in equilibrium with a liquid phase in the region $BaO{:}2TiO_2$. Recrystallisation from melts containing 63.7 mol TiO_2 by slow cooling in the temperature range 1370 to 1330^0C in an oxygen rich atmosphere can yield crystals of cm dimensions. The advantages of this process are:

1. TiO_2 is a non-contaminating component.

2. TiO_2 is relatively non-volatile and permits the use of open crucibles.

3. seeding of the melt, stirring and direct observation are permitted.

whilst the disadvantages are:

1. very slow growth rates with experimental runs lasting for up to 3 weeks during which time accidents occur viz power failure, loss of seed and furnace failures.

2. severe cracking due to the way the seed is held.

3. flux inclusions of the TiO_2 rich phase due to enhanced growth and insufficient rejection rate at the growth interface.

However, the method[4] has been successfully developed to give small quantities of crystals of device quality which are commercially available and expensive.

Other Perovskites

This Top Seeded Solution Growth (TSSG) method has been applied to a number of oxide ferroelectrics, Gunter[5] has reported interesting results with $KNbO_3$ which incongruently melts and requires solution growth from compositions containing excess K_2O. Good quality crystals have been obtained and are the basis of optical waveguiding in the range 0.5-1.3μm using the refractive index changes promoted by the E-O coefficient when modulated by 20 volts. This proposed structure offers the possibility of an integrated optical gate or switch, whilst the use of E-O and non-linear optical coefficients and favourable photorefractive properties make other devices possible with this technology.

$LiNbO_3$ and $LiTaO_3$ Structure Types

Both lithium niobate ($LiNbO_3$) and tantalate ($LiTaO_3$) belong to the point group 3m and have melting points of 1260^0C and 1650^0C and Curie temperatures of 1210 and 620^0C respectively. They possess a structure in which two of the three available sites are occupied by lithium. Although they possess similar electro-optic properties, with halfwave voltages of 3000V and 2800V respectively for $LiNbO_3$ and $LiTaO_3$, $LiTaO_3$ is not phase matchable for SHG. Furthermore, laser induced damage can be serious for $LiNbO_3$ and less so for $LiTaO_3$. However, minor chemical modifications to $LiNbO_3$ melts can inhibit laser damage at the phase matching temperature. $LiNbO_3$ is probably one of the most successful ferroelectric single crystal materials to be used for integrated and guided wave optics and related applications.

Growth of Lithium Niobate

It is over twenty years since the Czochralski melt growth method was extensively developed to pull crystals up to 3cm diameter; phase equilibrium diagrams were constructed in order to define the congruently melting composition essential for the growth of optical quality crystals from the melt. In the intervening years, a volume requirement for $LiNbO_3$ crystals has arisen from their use in SURFACE ACOUSTIC WAVE devices (TV filters), and this has motivated crystal growers to produce 10cm diameter crystals, Kg in weight whilst meeting cost and quality targets. Purer starting materials and the ability to produce such large crystals have been beneficial to the electro-optic field.

$LiNbO_3$ possesses good electro-optic and acoustic-optic properties, is transparent at wavelengths from the visible out to 5 or 6 microns and low loss waveguides may be constructed by titanium diffusion. The diffusion method gives low loss smooth profiles as compared with the alternative option of a ribbed (etched or milled) waveguide for operation at wavelengths of 0.83 to 1.55 microns.

There are some penalties and disadvantages in employing $LiNbO_3$ as an electro-optic material. Certain impurities which are difficult to eliminate give rise to the photorefractive effects at shorter wavelengths which are produced by photoexictation of the free carriers. However, in many applications, this can be engineered out by operating at longer wavelengths, but for others, it motivates the material scientist by defining a material need - high purity stoichiometric materials and crystals. The related ferroelectric compound Lithium Tantalate, $LiTaO_3$, has a lower photorefractive effect, smaller birefringence and higher laser damage threshold. It has the disadvantages of a lower Curies temperature 620^0C (cf 1200^0 for $LiNbO_3$) which is below the temperature necessary to effect titanium diffusion $1000+^0$C, and therefore would require repoling afterwards.

An alternative method to produce waveguides is to sputter epitaxial lithium niobate onto a $LiTaO_3$ substrate, but the method results in materials with high optical losses. A melting technique has also been attempted with little improvement, although there is considerable potential for liquid phase epitaxy (LPE) and MOCVD (metallo-organic chemical vapour deposition) techniques for the growth of $LiNbO_3$ onto $LiTaO_3$. In the LPE of $LiNbO_3/LiTaO_3$ a difference in refractive index of 0.01 is observed. A higher value can be effected by proton exchanged but leads to tighter confinement and a need for high E-O efficiencies. A compromise emerges between engineering around the problem of efficient coupling of the light onto the guide and electro-optic efficiency.

Ferroelectric Oxide Tungsten Bronzes

The ferroelectric oxides exhibiting the tungsten bronze stucture type have been extensively researched for nearly two decades because of their considerable potential in electro-optic, pyroelectric and piezoelectric devices. The tetragonal tungsten bronze unit cell[6], in the projection of the (001) plane, consists of ten distorted BO_6 octahedra sharing corners in such a way that five different sites designated A_1, A_2, c, B and B_2 are available for cations:

4 A_1 sites in pentagonal tunnels parallel to 'c' axis surrounded by 15 oxygens.
2 A_2 sites in tetragonal tunnels parallel to 'c' axis surrounded by 12 oxygens.
4 C sites in trigonal tunnels parallel to 'c' axis surrounded by 9 oxygens.
2 B_1 sites located at the centre of the rectangular faces.
8 B_2 sites surrounding the A_2 sites.

The general formula $(A_1)_4(A_2)_2C_4(B_1)_2(B_2)_8O_{30}$ describes a unit cell for which a wide variety of substitutions can be made in the available cation sites, provided that overall electrical neurtrality is retained. The polar axis is the $\leqslant 001 \geqslant$ c direction for all cations excepting Pb and Bi. Ferroelectricity is found in this system for partially filled structures (vacant A_1, A_2 and C sites), filled structures (vacant C sites) and fully filled (no vacancies).

Crystal Growth

Neurgaononkar's team at Rockwell has made important progress in the growth of bulk crystals and epitaxial films of selected compositions which offer generous electro-optic coefficients, often more than an order of magnitude greater than $LiNbO_3$. Most crystals have been grown from the melt (Czochlralski) method. The phase equilibria diagrams for many of the preferred tungsten bronze niobates are complex and congruency in melting is not often found for ferroelectric members. Barium strontium niobate $(Ba_xSr_{1-x})_5Nb_{10}O_{30}$ has been extensively studied for values of $0.75 \geqslant x \geqslant 0.25$. It is a partially filled bronze with five cation vacancies per unit cell and can be viewed as a solid solution of $xBaNb_2O_6$ - (1- x) $SrNb_2O_6$. This gives rise to numerous problems of chemical homogeneity, susceptibility to laser damage and AC dielectric losses. Crystal twinning can also occur. A filled bronze $Ba_4Na_2Nb_{10}O_{30}$ is grown from near stoichiometric melts at 1500^{0}C and on cooling passes through its Curie point of 560°C into the polar tetragonal phase. It slightly distorts below 260°C to an orthorhomibic form where the face diagonal of the tetragonal cell becomes the reference axis and its optic axis lies in the a-c plane. The polar axis is 'c' and twinning is subsequently removed by cooling from around 300°C with stress along one of the tetragonal 'a' axes. It is an excellent, highly efficient material for SHG at 1.06μm and appears not to suffer from optical damage.

The fully filled bronze $K_6Li_4Nb_{10}O_{30}$ can only be grown by a modified solution melt technique. There exist a large range of compositions exhibiting the bronze structure[7] of which only those with c ⩾4.02 exhibit ferroelectricity. The complex phase equilibria of these materials has hindered progress with what should be very interesting materials for device structures.

One of the overriding problems with the tungsten bronze structure is that it is very forgiving and retained for a wide range of cation occupancies and compound stoichiometries. In partially filled $Ba_{1-x}Sr_xNb_2O_6$ and $K_2Sr_9Nb_{20}O_{60}$[8] there are 5-10% ionic vacancies leading to distorted NbO_6 octahedra. Fully poled domain configurations are unstable and drift with time; this ageing is accompanied by both increasing and decreasing loss with time depending upon the lift history of the crystal and its poling conditions. Large changes in dielectric constant occur during poling which have not yet been fully explained.

Layer Structure Compounds $(Bi_2O_2)^{2+}(A_{1-x}B_xO_{3x+1})^{2-}$

A large number of layer structure compounds with general formula $(Bi_2O_2)^{2+}(A_{1-x}B_xO_{3x+1})^{2-}$ have been reported (Smolenski et al[9], Subbarao[10]), where A = Ca, Sr, Ba, Pb, etc; B = Ti, Nb, Ta and x = 2, 3, 4 or 5. The structures were first described by Ausivillius[11] in 1949 in terms of alternate $(Bi_2O_2)^{2+}$ layers and perovskite layers of oxygen octahedra. Only bismuth titanate $Bi_4Ti_3O_{12}$ $[(Bi_2O_2)^{2+}(Bi_2Ti_3O_{10})^{2-}]$ has been studied in detail in the single crsytal form[12]. It was studied for optical storage applications (Cummins 1967) because of its unique ferroelectric optical switching properties. There is a revival of interest in the switching of the optical indicatrix as a means of making an optical switch. The two prime candidate ferroelectrics are $Gd_2(MoO_4)_3$ and $Bi_4Ti_3O_{12}$, the latter being preferred for a number of reasons. It is a monoclinic (pseudo-orthorhombic) ferroelectric[8] with its spontaneous polarisation lying in the a-c plane consisting of a large component (~50μCcm^{-2}) parallel to 'a' and a smaller one (~4μCcm^{-2}) parallel to 'c', see Figure 2.

Each can be switched independently, P_a by a coercise field of $50kVcm^{-1}$ and P_c by one of $4kVcm^{-1}$. However, by switching P_c, the optical indicatrix[13] is reflected in the a-b axis, such that when viewed down the 'b' axis, it appears to have rotated by 58^0. If the crystal thickness is chosen to give π retardation, then an electro-optic shutter is available for polarised light. Since switching inverts the c axis, the only lattice parameter to change is the monoclinic angle and as it is nearly 90^0[14], there is no significant change and clamping effects avoided. Fast switching has been demonstrated at 100ns for a field of $18kVcm^{-1}$. The material exhibits memory and accurate voltage control is not required to achieve π retardation and it operates through the optical indicatrix. Furthermore, its low dielectric constant and high insulation resistance commends it for electro-optic modulators.

Growth of Bismith Titanate

The compound $Bi_4Ti_3O_{12}$ incongruently melts losing Bi_2O_3 and therefore solution growth techniques have had to be developed[15] for the provision of single crystals. As with $BaTiO_3$, the phase diagram reveals a region where tetragonal (high temperature phase) $Bi_4Ti_3O_{12}$ is in equilibrium with a $Bi_{12}O_3$ rich phase. Again, a non-contaminating flux is available.

Crystals are grown by slow cooling melts in sealed platinum crucibles in the range $1100\text{-}900^0C$. Crystals spontaneously nucleate and if the conditions are carefully controlled, large sheets (5cm x 5cm x 0.5cm), covering the melt surface are obtainted. Corresponding to the a, b and c directions. They may be twinned and left to be poled in the 'a' direction (Hopkins and Miller) to obtain a uniform optical response. The natural growth habit which favours ab growth in the $Bi_2O_3\text{-}TiO_2$ system is difficult to modify and attempts employing flux pumps designed to encourage 'c' axis growth have not been successful (P M Dryburgh 1968). However, these crystals provided considerable interest for device studies[16]. Since the crystals have large a-b planes, whereas a-c planes are required, epitaxial studies have been made. Francombe[17] has reviewed the growth of single orientation [010] films onto MgO and $MgAl_2O_4$ single crsytal substrates. He and his co-workers used an RF sputtering technique[18] and films microns thick can be grown under optimum conditions with a target composition of $0.8Bi_4Ti_3O_{12} + 0.2Bi_{12}TiO_{20}$.

Other Crystalline Forms

It is seen that many of the ferroelectric single crystals required for non-linear optics are difficult to grow and are not commercially available, the only exceptions being $LiNbO_3$ and $LiTaO_3$. Therefore alternative methods for obtaining optical quality ferroelectric oxides have been sought. A number of perovskite ferroelectrics[14] can be made as non-porous large diameter transparent ceramics, and of the systems studied[19,20], the lead lanthanum zirconate system $Pb_{1-3x/2}La_xZr_{1-y}Ti_yO_3$ has been extensively developed because of its versatility and accessibility as a high grade ceramic material. As compared with the single crystal cases described earlier, solid solution effects and chemical substitutions can be employed to tailor the composition to the desired properties. This is the major benefit of being able to carry solid state reactions to form our optical material. The PLZT system offers a wide range of optical /ferroelectric behaviour which has been well reviewed by G Haertling[21]. The phase diagram of PLZT can be divided into four major regions: (1) tetragonal linear electro-optic ceramics, (2) rhombohedral ceramics with memory and display potential, (3) pseudo-cubic quadratic electro-optic (Kerr Effect) and (4) anti-ferroelectric-ferroelectric switching for light scattering. Fine grain ceramic structures ($G \leq 3\mu m$) are normally used for devices using the true electro-optic effects where as coarse grain structures ($G \geq 10\mu m$) may be employed in light scattering devices. There are extensive reports on their uses and applications, including (a) photosensitivity and imaging characteristics of ion implanted PLZT[22], (b) displays with memory[23], (c) page composers and optical shutters for video projection[24], and (d) 3D viewing shutters[25]. Spatial light modulators for optical computing employing either integrated PLZT-silicon structures or flip chip solder bump bonded structures combine the light sensing speed of silicon with the large optical non-linearity of PLZT[26].

Only selected systems can be made as high quality optical ceramics of which the perovskites are dominant. It has been shown that other structure types viz tungsten bronzes and layer structure components have attractive non-linear behaviour so alternative methods are being investigated for fabricating ferroelectric electro-optic devices. Wet chemical methods are used for preparing homogeneous stoichiometric powders for the fabrication of a high performance unifom ceramic bodies[27]. These solutions can be selectively used for the preparation of sol-gels which in turn can be deposited as thin films onto suitable substrates. Although several deposition and thermal treatments are required to build a film of microns thickness, optical quality ferroelectic films are emerging[28], which have similar properties to the bulk ceramic material. It appears to be a fertile field for the provision of thin ferroelectric films. Other methods being actively pursued include physical deposition (sputtering, etc), chemical vapour deposition and liquid phase epitaxy. Of these, MOCVD looks very promising with the prospect of low temperature growth through photolytic reactions. Trundle et al[29] are designing precursors with selected absorption bands, high extinction coefficients, thermal stability and similar volatilities.

CONCLUSIONS

It has been shown that ferroelectric oxides containing the BO_6 oxygen octahedral building block offer interesting and desirable non-linear optical effects. The compounds of major interest are not yet commercially available and offer formidable problems to the crystal grower. The most successful crystals which are readily available in size and quality are $LiNbO_3$ and $LiTaO_3$. There has been significant progress with the growth of certain tungsten bronze oxide ferroelectrics for optical components but their future must be in some doubt until some of the fundamental underlying problems of vacancies, ageing and AC losses have been resolved.

Electro-optic ceramics have many virtues and are finding more applications. However, the most exciting future for the materials scientist, device physicist, applications engineer and the accountants will be the growth of epitaxial films onto readily available substrates so that the concepts of full integration may be realised.

REFERENCES

1. M DiDomenico Jr & S H Wemple, Appl Phy Lett, 12, 352-   , 1968.

2. M DiDomenico Jr & S H Wemple, J Appl Phys, 40, 720 -   , 1969.

3. D Rase & R Roy, J Am Ceram Soc, 38, 110 -   , 1955.

4. A von Hippel, Final Report, March 1963, AFCRL - 63 - 92. Contract No, etc.

5. P Gunter, Electro-Optical Effects in Dielectric Crystals, Ferroelectrics 75, 5-23, 1987.

6. M H Francombe & B Lewis, Acta Cryst 11, 696- , 1958.

7. F W Ainger, J A Beswick, W P Bickley, R Clarke, G V Smith, Ferroelectics in the Li_2O-$K_2ONb_2O_5$ System, Ferroelectrics 2, 183-199, 1971.

8. F W Ainger, J A Beswick, S G Porter & R Clarke, Feroelectric in the K_2O-SrO-Nb_2O_5 System, Ferroelectrics 3, 321-325, 1972.

9. G A Smolenski, V A Isupo & A I Agranovskaya, Fiz Tverdogo, Tela, 1, 169-170, (1959).

10. E C Subbarao, A Family of Ferroelectric Bismuth Compounds, J Phys Chem Solids, 23, 665-676, 1962.

11. B Aurivillius, Arkiv Chemi 1, 463 (1949) ad 2, 519 (1950).

12. S E Cummins & L E Cross, Appl Phys Lett, 10, 14-16, 1967.

13. S E cummins & L E Cross, Electrical and Optical Properties of $Bi_4Ti_3O_{12}$ Single Crystals, J Appl Phys, 39, 2268, 1968.

14. R E Newnham, R W Wolfe & J F Dorian, Mat Res Bull, 6, 1029-1040 (1971).

15. L G Van Uitert & L Egerton, Bismuth Titanate, A Ferroelectric J Appl Phys, 32, 959, 1961.

16. G W Taylor & W F Kosonocky, Ferroelectric Light Valve Array for Opical Memories, Ferroelectics, 3, 81-99, 1972.

17. M H Francombe, Ferroelectic Films and Their Device Applications, Thin Solid Films, 13, 413-433, 1972.

18. S Y Wu, W J Takei, M H Francombe & S E Cummins, Domain Structure and Polarisation Reversal in Films by Ferroelectric Bismuth Titanate, Ferroelectrics, 3, 217-224, 1972.

19. G Haertling and C Land, Hot Pressed (PbLa)(ZrTi)O_3 Ferro-electric Ceramics for Electro-Optic Applications, J Am Ceram Soc, 54, 1-11, 1971.

20. A Krummins, Specific Solid State Features of Transparent Ferroelectric Ceramics, Ferroelectrics, 69, 1-6, 1986.

21. G Haertling, PLZT Electro-Optic Materials and Applications - A Review, Ferroelectrics, 75, 25-55, 1987.

22. P S Peercy & C E Land, Photographic Image Storage in Ion Implanted PLZT Ceramics, Nuclear Instruments & Methods, 209-210, 1167-1178, 1983.

23. K L Bye, High Birefringence PLZT Materials for Low Voltage Displays, Ferroelectrics, 10, 29-33, 1976.

24. W E Perry and B M Soltoff, Television Applications of PLZT Ceramics, Ferroelectrics, 10, 201-205, 1976.

25. J A Roese & A S Khalafalla, Stereoscopic Viewing with PLZT Ceramics, Ferroelectrics, 10, 47-51, 1976.

26. C J Kirkby, M J Goodwin, A D Parsons, W J Stewart & I Bennion 8 x 8 Element Hybridised PLZT/Silicon Spatial Light Modulator Elec Lett, 25/18, 1260-61, 1989.

27. A Patel, M T Goosey, R W Whatmore & F W Ainger, Sci Ceram, 14, 237-41, 1988.

28. A Patel (Plessey Research Technology Ltd) Private Communication.

29. C Trundle & C J Brierley, Precursors for Thin Film Oxides by Photo-MOCVD, Applied Surface Science, 36, 102-118, 1989.

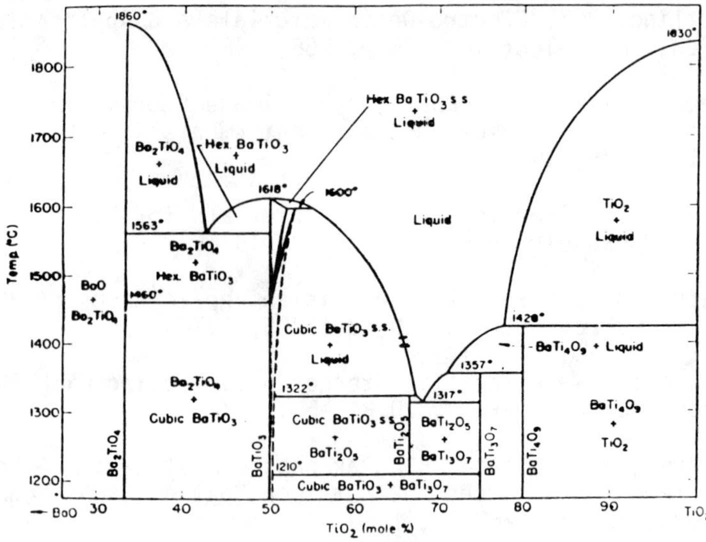

FIG. 1 PHASE EQUILIBRIUM DIAGRAM OF THE SYSTEM $BaO-TiO_2$ (REF. 3)

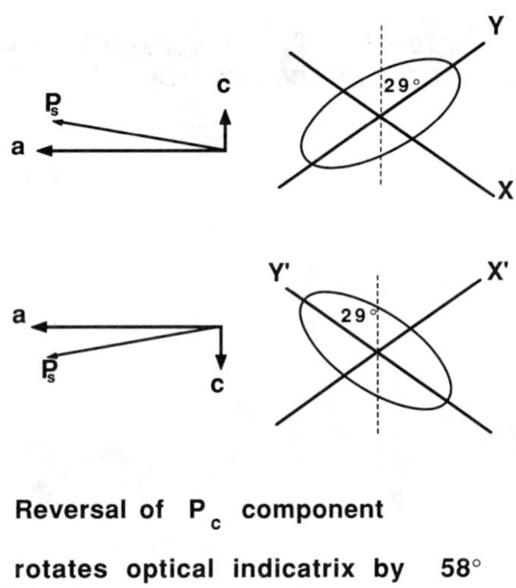

FIG. 2

PHOTOELECTRONIC EFFECTS IN PLZT CERAMICS

Gene H. Haertling
Clemson University
Clemson, SC 29634-0907

ABSTRACT

The current trend in the electronics industry toward increasing miniaturization and greater integration has placed more stringent demands on the ceramic materials used in electronic and electrooptic applications. In regard to the electrooptic materials, these increased requirements have resulted in (1) higher operating electric fields and dielectric stress levels, (2) greater exposure to high intensity UV and visible light during operation, (3) higher speed of operation and (4) reduced element size (i.e., higher resolution). From these, it is obvious that the behavior of the electrooptic ceramic materials under combined stress levels is vitally important. This study reports on various photoelectronic effects which are known to occur in the PLZT family of ceramic electrooptic materials such as photorefraction, photoconduction, photoassisted domain switching, photoexcited space charge and photovoltaic effects. Shutter contrast ratio results and UV-generated space charge behavior are presented and described in terms of shutter performance.

I. INTRODUCTION

Since their development in 1969, PLZT ceramics have commanded a considerable amount of attention from research investigators throughout the world, partly because of their promising application to such devices as shutters, displays, image stores and holographic memories and partly because of their numerous interactive properties which provide fertile ground for many unusual and fascinating effects.[1,2] These materials are, by nature, ferroelectric (FE) or antiferroelectric (AFE) at room temperature and consist of a series of solid solutions involving both lead zirconate and lead titanate with varying amounts of lanthanum oxide. As ceramic solid solutions, their chemical compositon and other properties can be varied continuously from one end member composition to the other, achieving both optimum behavior and high reproducibility. The phase diagram of the PLZT system is given in Figure 1.

The PLZT materials are especially desirable from a research standpoint because, like all ferroelectrics, they possess dielectric, piezoelectric and pyroelectric properties; but unlike most FE ceramics, they are also optically transparent and photosensitive, thus opening up a whole new and larger dimension of investigative study for these materials. These photosensitive properties include (1) photoconductivity, (2) photovoltaic properties, (3) photo-assisted domain switching, (4) enhanced photo-assisted domain switching via ion implantation, (5) photochromic effects, (6) photomechanical (photostrictive) behavior, (7) photorefraction and (8) photoexcited space charge phenomena.[3-9] While materials with such properties as these hold promise for many new applications, it should also be remembered that these same photoelectronic properties can, and often do, lead to degradation and fatigue.

The study of such optical properties and related optical phenomena is, of course, not new for FE single crystals. Research is this area has been conducted for many decades and is presently providing valuable insight into the phenomena observed in ceramic materials. While the analogies of behavior between single crystals and ceramics are generally valid and applicable, it should be pointed out that ceramics are uniquely different from single crystals in several ways, especially in regard to their polycrystalline nature (grain size effects), their random crystalline orientation and the presence and make up of grain boundaries; and hence, they often possess phenomena not found in single crystals. Moreover, the recent trend toward miniaturization and integration involving thin film materials has also borne out the fact that the electrical and optical properties of these structures are, once again, different from the bulk single crystal or ceramic material.[10] Thus, as the size of these structures becomes smaller (i.e., nanostructures), the stress levels increase and new effects and behavior are to be expected. Under these circumstances, grain size, grain boundary and grain orientation effects which usually influence ceramic behavior are anticipated to become even more significant.

It is the object of this paper to briefly describe the various photoelectronic effects which are operative in the ferroelectric PLZT ceramic materials and have been reported by previous investigators. Furthermore, new results are described which provide additional evidence that these materials possess highly complicated physical and electronic structures and are extremely sensitive to optical stimulii which can affect performance.

II. MATERIALS

The optical, electrooptic and other photo-related properties of the PLZT materials are directly linked to the chemical, structural and electronic condition of the materials as they are formulated and fabricated. Consisting of homogeneous solid solutions of $PbZrO_3$ and $PbTiO_3$ modified with substantial amounts of La_2O_3 within a solid solubility range of 0-30 atomic percent, these materials possess, among other attributes, reduced Curie points and significantly enhanced optical transparency (greater than 97% neglecting reflection losses). The higher valent La^{3+} substituting into the Pb^{2+} position of the ABO_3 unit cell acts as an electron donor, supplying excess electrons which are compensated in the PLZT lattice through creation of vacancies on both the A and B sites of the ABO_3 perovskite structure. The existence of these vacancies and the ease with which they are formed (due to the volatility of the PbO) has a profound effect on the behavior of the material, being credited with such effects as (1) enhanced sintering, (2) automatic electronic compensation for off-valent additives, (3) easier ferroelectric domain switching and (4) enhanced photosensitivity. Further implications of such a structure are pointed out by Krumins and co-workers[5] as it relates to transparency in these highly doped and compensated materials.

Compositions in the PLZT system which possess a reduced Curie point near room temperature constitute a special group of materials known as relaxors. They are so named because of their dielectric relaxation characteristics as a function of frequency and the dispersion of their Curie point over a rather large temperature range. Also known as diffuse phase transition (DPT) materials, these relaxors have intrigued many investigators attempting to define the nature of the near-ferroelectric (pseudo-cubic) state separating a definite FE phase from a clearly cubic (non-FE) state.[11-17] This phase has been given several names including slim-loop FE, quasi-FE, α phase, penferroelectric, γ phase, glassy polarization and quasi-paraelectric. Pioneered by Smolensky[18] and further elucidated by Cross [19], these DPT materials are generally believed to consist of microscopic polar regions characterized by compositional fluctuations leading to a local symmetry condition (polar) which is lower than the overall macroscopic symmetry (cubic). Although their true nature has not yet been definitely established, it is clear that they constitute an important category of materials because from within this group come the highly sensitive, non-linear electrooptic ceramics typified by such compositions as 9/65/35 (La/Zr/Ti) and 8/70/30. They are located in the cross hatched region of Figure 1.

The DPT materials possess, for all practical purposes, cubic symmetry at room temperatue; and as a consequence must be subjected to an external stimulus in order to make them optically active. An external electrical field is one means commonly used to impart such activity, leading to various polarization-related properties; e.g., light scattering, electrooptic birefringence, photoconductivity, photorefraction and space charge effects. Being of inherently cubic symmetry, these materials retain these properties only as long as the electrical field is applied and will relax back to a non-active state when it is removed.

The FE materials, on the other hand, possess true memory, and when once subjected to an electric field sufficiently high to switch the internal polarization (poling), they retain their activity indefinitely, providing the polarization is not changed through some other external means. Compositons of importance here include both the FE rhombohedral (7/65/35, 8/65/35) and the FE tetragonal (8/40/60, 12/40/60) phases.

A third class of materials in the PLZT systen, the antiferroelectric group, are also under study for their interesting electrooptic properties.[20-21] In this case, the AFE phase is the non-polar state, and like the DPT materials, it is electrically enforced into a FE state which exists only so long as the field is applied. Slight compositional changes at the FE-AFE phase boundary will produce a material containing a "mixture" of FE and AFE phases. This composition (7.6/70/30) is very temperature sensitive but possesses the desirable characteristics of a polar memory FE state combined with an electrically switchable capability to another memory, highly tramsparent, non-polar AFE state.

III. PHOTOELECTRONIC PROPERTIES

As mentioned previously, the FE memory materials must be initially poled in order to make them electrically and optically active. Once poled, they then possess a number of useful electrical and optical properties; however, it has been demonstrated that these properties are quite sensitive to other influences such as stress (mechanical, electrical), temperature and high intensity optical or UV radiation. Furthermore, combined influences such as temperature and high electric field or electric field and high intensity radiation are particularly detrimental. These same effects, in almost all cases, are also found in the DPT relaxor materials, but it should be remembered, in this instance, that an electrical bias must be maintained on the sample in order to realize a response.

The following photo-related phenomena are the major effects so far observed in FE or electrically-induced FE materials. In most cases, the responses occur as a result of electromagnetic radiation with wavelengths across the optical range and including the near-UV and near-IR; i.e., 360nm to 1000nm.

1. Electrooptic Properties

Birefringence - Transverse Ferroelectric PLZT ceramics display optically uniaxial properties on a microscopic scale and also on a macroscopic scale when polarized with an electric field. In uniaxial crystals there is one unique symmetry axis, the optic axis (co-linear with the polarizaton vector in ceramic PLZT), which possesses different optical properties than the other two orthogonal axes; i.e., light traveling along the optic axis has a different index of refraction (n_o) than light traveling in a direction perpendicular to the optic axis (n_e). The absolute difference between the two indices is known as the birefringence (Δn), and it is this quantity which is influenced by the electric field. Birefringence is also a measure of the phase difference between the two orthogonal light waves traveling through the ceramic. It can be translated into light amplitude modulation, as in a light gate, through the use of linear polarizers.

The DPT materials (9/65/35) are most useful in the birefringent mode of operation because they possess the desirable property of self-relaxation back to a non-birefringent off condition when the electric field is removed. Since this mode requires the generation of a transverse electric field for maximum effect, interdigital surface electrodes (Cr/Au, Al) are commonly employed. This type of electrode not only produces a transverse field, but it also lowers the absolute operating voltage, as well. Its main disadvantage, however, is that it produces non-uniform electric fields across the gap between the positive and negative electrodes, and this leads to degradation in performance. As a result, surface slotted electrodes which produce a more uniform and effective transverse field at an even lower operating voltage is preferred. A typical configuration for observing this effect (I vs. E) is given in Figure 2.

<u>Scattering - Longitudinal</u> Scattering occurs in all of the electrooptic PLZT ceramic materials to some degree, but it is most prominent in the rhombohedral, memory ferroelectrics such as 7/65/35 which possess higher birefringence. The high birefringence is the source of the higher light scattering brought about as a result of refractive index discontinuities at the domain walls. Grain boundaries and porosity are also sources of light scattering, but are of lesser importance and not electrically variable.

The means of achieving optical contrast via scattering in these materials is through the use of controlled switching of domains. Light is preferentially scattered along the polar direction at a low angle, and consequently, transverse domains appear darker than longitudinal domains. Transparent electrodes are necessary to observe the varying domain orientations. Polarizers are not employed. A typical setup for studying this effect is also given in Figure 2.

2. Photovoltaic Effects

The photovoltaic effect is the creation of a steady state electromotive force or voltage which is generated when the incident radiation is of sufficiently high energy to create free carriers which move preferentially in one direction in the absence of an external field. All FE materials, being pyroelectric with a spontaneous internal polarizaton, display this effect providing the electrical resistivity of the material is high enough to prevent total internal compensation. As pointed out by Brody[3], the open circuit photovoltages can be very high, exceeding 1500 volts across a one centimeter long, electroded sample of PLZT 7/65/35. The magnitude of the effect was found to be dependent on composition (polarization), grain size, electrode gap, temperature, wavelength of radiation and radiation intensity. A typical setup for observing photovoltaic behavior is shown in Figure 3.

According to a model developed for these materials, larger than bandgap (3.35eV for PLZT) steady-state photovoltages are possible even though a less-than-bandgap voltage appears across each grain (when illuminated) as a result of the series addition of individual grain voltages. The sign of the photovoltage always has the same polarity as the poling voltage.

3. Photoconductivity

The generation of electrons in the conduction band or holes in the valence band of a FE material via illumination and subsequent current flow under an applied field defines the material as a photoconductor. Photoconduction is intimately associated with the photovoltaic effect and can be observed by monitoring the flow of current under short circuit conditions or through an external resistor connected across the electrodes of the poled FE. The photocurrent behaves as if it were generated by a constant photovoltage in series with the external resistor and/or the illuminated resistance of the FE. Brody[3] has reported a short circuit photocurrent of 0.31 uA/W/cm² for 7/65/35. This value compares favorably with a value of 0.24 uA/W/cm² obtained in our laboratory with a similar composition. A lower value of 4 nA/W/cm² was reported by Uchino, et. al.[4] for $PbTiO_3$-based ceramics which characteristically possess lower switchable polarization.

For the evaluation of the photocurrents in DPT materials such as 9/65/35, an external voltage source can be placed in series with the sample. In this situation, the photocurrent generated will be dependent upon the bias voltage as well as the radiation wavelength and intensity. As a function of temperature, it has been observed that the photovoltage decreases with increasing temperature while photocurrent generally goes through a maximum. Both disappear at the Curie point.

4. Photoassisted Domain Switching

PLZT ceramics may be described as photoferroelectric (PFE) because they are both photosensitive and ferroelectric. As a photosensitive material, they are readily influenced by UV and visible light which photoexcite free electrons within the crystalline structure creating space charge that thermally drifts through the lattice or moves under the influence of an electric field. As a FE, they can be polarized by switching or reorienting their internal polarization with an electric field. When both of these processes take place at the same time, a phenomenon known as photoassisted domain switching (PDS) occurs.[7] This is demonstrated in Figure 4 which shows the actual shift in the hysteresis loop (lowering of E_c in the bias direction) for 7/65/35 when UV is applied under voltage bias.

Although applicable to switching in general, PDS is usually thought of as the process by which spatially selective switching is used to produce images or holographic interference patterns in the material. The three major processes which contribute to PDS include

(1) photoexcitation of carriers both from trapping centers in the PLZT bandgap and across the bandgap by UV light of energy equal to or greater than the bandgap of 3.35 eV (visible light can also photoexcite carriers from trapping centers in the bandgap but with less efficiency), (2) carriers photoexcited to the conduction state thermally drift or move under the influence of an applied field to new trapping sites beyond the absorption depth of the light which establish a space charge field that modulates the applied field and aids in the domain switching process by effectively lowering the switching field (E_c) of the material and (3) photoexcited carriers remaining in the conduction state contribute to a steady state photocurrent which is driven by the bulk photovoltaic effect. An example of an image stored in PLZT 7/65/35 by this process is shown in Figure 5.

5. Ion Implantation Enhanced PDS

The various studies on PDS in PLZT all point out that the most important mechanisms contributing to the PFE image storage capabilities of PLZT is the space charge which is stored in the UV absorption region near the surface of the material rather than in the bulk. This concept is supported by the work of Micheron, et.al.,[22] and Burgess and coworkers[23] which indicate that the photosensitivity of PLZT is not noticeably enhanced as a result of bulk modifications produced by doping with off-valent additives. Thus, the key to controlling space charge and photosensitivity appears to be the material surface. This reasoning led Land and Peercy[7] to concentrate on a surface modification technique known as ion implantation in order to increase the image storage sensitivity through enhanced PDS.

In this technique, ions such as H, He, Ar, Ne, Al, Cr and Ni are implanted into the surface of the PLZT at ion energies ranging from 200-500 KeV, producing a disordered layer about one micron thick. Since the higher photosensitivities are obtained with the heavier ions, the domain mechanism for increased photosensitivity is believed to be associated with implantation disorder (damage) rather than chemical effects. This disorder results in significantly reduced dielectric constant, dark conductivity and photoconductivity of the implanted layer. Using this technique, photosensitivities in 7/65/35 have been increased by several orders of magnitude (100 ergs/cm^2). This makes it the most sensitive, non-volatile, selectively erasable, image storage medium available today; however, it is still 3 to 4 orders of magnitude less sensitive than silver halide photographic film.

6. Photorefraction

As previously mentioned under photoassisted domain switching, exposing PLZT to UV light with a wavelength (λ = 365nm) near the absorption edge of PLZT (λ = 370nm or 3.35 eV) results in the excitation of electrons or holes from lower energy traps into the conduction band at a rate dependent on the local light intensity. These carriers thermally diffuse or drift under the influence of an electric field and are subsequently re-trapped at regions of low illumination. The resultant space charge set up by the trapped carriers will alter the switching field of a FE material (as in PDS), however, this same space charge will also create or alter the refractive index and birefringence of the material via the electrooptic effect.[24] This phenomenon is known as photorefraction.

Photorefraction is present in all ferroelectrics to some degree, but is very noticeable in the highly transparent, non-memory PLZT ceramics such as 9/65/35 and 8/70/30. Their high sensitivity is, first of all, due to the fact that birefringence is much easier to detect in these non-memory materials since they should possess no optical activity in their off state; secondly, the electric field must remain on continuously during the open state of the shutter which can lead to additional space charge generation as a function of time; thirdly, the common practice of using surface, interdigital, transverse field electrodes leads to non-uniform electric fields which are especially high ($>$ 50kV/cm) immediately adjacent to the electrode edges and fourthly, while under the influence of an electric field, these DPT materials are in an enforced FE state which has a time dependent inclination toward a memory condition when the field is removed. As a result, the photoinduced change in refractive index (PCI) effect, or remanent light flow (RLF) as it is sometimes called, is a serious concern in many electrooptic devices and often limits their performance and applications.

A considerable amount of study and attention has been given to the photorefractive effect by various investigators.[22-28] It will be discussed more thoroughly in the remaining portions of this paper.

IV. EXPERIMENTAL

Hot pressed, polished and electroded samples of PLZT compositions 9/65/35, 8/70/30 and 7/65/35 were selected for further study of their photorefractive behavior. Single sided, interdigital, surface electrodes consisting of vacuum deposited Cr/Au, sputtered ITO or mechanically slotted and chemically plated Ni/Cu were applied to the samples after annealing at 600°C for 1 hour. Electrical, electro-optic, photoconductivity and RLF measurements were made on the samples as a function of electric field, temperature, time and light exposure. A mercury arc lamp (15mW/cm^2) was used as the high energy light source in conjunction with various common filter materials ranging in wavelength from 360 nm to about one micron.

The setup utilized in the observations of the photo-electronic effects is shown in Figure 6. The basic unit consisted of a Leitz polarizing microscope equipped with a heated stage and long focal length objectives providing magnification from x50 to x300. The sample was mounted in the optical path of the microscope between polarizers with sufficient space allowed for impingement of the UV light rays at an angle of approximately 30° and a distance of 20 cm. A dc hysteresis looper was connected to the sample through a 22 Mohm resistor used for monitoring photo-currents. Hysteresis loops (polarization vs. electric field) were plotted on an X-Y recorder which served also to record light intensity I vs. E by means of a United Technology, model S370, photometer.

A typical evaluation run consisted of obtaining a thermally depoled hysteresis loop and a I vs. E loop at room temperature, and then subjecting the specimen to various levels of UV radiation for specified periods of time and different voltages of opposing polarity.

After observing the various RLF effects and recording contrast ratios, the temperature was raised to a new value and the measurements repeated. Selected photoconductivity and time dependent space charge measurements were made at various temperatures with the majority of them being conducted at 100°C or greater where the higher photocurrents were obtained.

The various filters selected for evaluation of PLZT photocurrent response consisted of those materials normally used in conjunction with the preparation and fabrication of electrooptic devices; i.e., ruby lith (red), UV filter (blue), soda lime glass, KG-3 heat absorbing glass, indium tin oxide (ITO) conducting electrode film, HN-32 polarizer and PLZT (.375 mm thick) polished plate. Their respective spectral responses are given in Figure 7.

V. RESULTS AND DISCUSSION

A. Photoconductivity Effects

PLZT ceramics are well known to be highly sensitive to UV radiation at wavelengths shorter than 370 nm since their absorption in this region becomes extremely high. High intensity UV light of this wavelength is especially effective in producing photocurrents in PLZT at high temperatures where the mobility of the photoexcited carriers is higher. A simple demonstration of this behavior is shown in Figure 8. In this figure, the UV stimulated photocurrent response as measured by a voltage across the 22M ohm resistor in Figure 6, is recorded as a function of various filter materials inserted between the unfiltered mercury arc UV source and the PLZT sample. It can be seen that the metal stop and the ruby red material are the most effective in absorbing the incident radiation and thus lowering the photocurrent. This is closely followed by the PLZT plate which would be expected as a result of its absorption edge being slightly higher in wavelength than the peak output line of the UV source (365 nm). Other strong lines of the UV lamp (546 and 578 nm) are not filtered by the PLZT plate and are free to contribute to the photoconductivity. Since the PLZT plate produces such a large effect, it is evident that the mid to high visible wavelengths contribute little to the photoconductivity in PLZT. The UV blue filter is observed to be only moderately effective in reducing photoconductivity because it is normally designed to pass the critical wavelength region of 360-380 nm. Both of the glasses have the least influence on photoconductivity primarily because they are transparent throughout the wavelengths of interest.

Further examination of the photoconductivity in PLZT 9/65/35 is given in Figure 9. Shown in this figure is the response time of the photoconduction process which is considered to be approximately one second for complete response in this material at 100°C. Also shown is the effect of bias voltage on the generated photocurrent as a function of both modes of operation; i.e., transverse and longitudinal. Results here point out that an approximate linear relationship exists between the photocurrent and bias voltage; and furthermore, this relationship is similar for both modes of operation.

B. Space Charge/Photorefractive Effects

As fabricated, PLZT electrooptic shutters and displays exhibit a highly non-linear quadratic (Kerr effect) and somewhat hysteretic optical output as a function of voltage as seen in Figure 10(A). This non-linearity occurs in spite of the rather linear behavior of the material with respect to its hysteresis loop (Q vs. V) shown in Figure 10(B). The excellent symmetry of the optical loop (I vs. V) demonstrates a well annealed material with little or no mechanical or electrical bias (stress, space charge) as fabricated.

This desirable condition may be contrasted with the results in Figures 11 and 12 obtained for the same composition but subjected to a 5 minute UV exposure while under an electric field bias of 40 kV/cm (100 V/mil). As can be seen, for each temperature, the thermally depoled and symmetrical loop on the left is transformed into a highly non-symmetrical loop which depends significantly on the direction of the electrical bias, the temperature of the measurement, and perhaps, the type of electrode. It should be noted in Figure 11 (ITO interdigital electrodes) that the bias polarity sensitivity is the same at both low and high temperatures even though the output intensity is reduced at the higher temperature, as expected. This is not the case in Figure 12 (Cr/Au electrodes) where we observe a change in behavior from room temperature (25°C) to 56°C and higher. At 25°C, the UV/bias exposure treatment favors a higher optical output in the direction of the bias; however, this effect gives way to a completely opposite behavior at about 56°C where the exposure drastically lowers the output in the direction of the bias. The effect at 100°C is similar in behavior but faster, sometimes occurring in a matter of seconds.

As explained previously, a large part of this high temperature effect is due to UV photoexcited space charge which moves under the influence of the electric field and creates an opposing field that nullifies the externally applied field; hence, the lowering of the output. Fortunately, with a combination of time and no field, this detrimental effect is completely erased. The room temperature behavior, however, is not yet well understood for this DPT material and could consist of two different mechanisms; i.e., (1) a simple process of the UV assisting the further field enforcement of a FE phase from the non-FE phase, thus increasing birefringence and output or (2) a UV-assisted space charge generation as in the high temperature case, but with carriers of opposite sign (holes vs. electrons). It is possible that the different types of electrodes or the variation in batches of material could cause this varying behavior, but one thing is abundantly clear, that further work needs to be done to clarify the situation.

Figure 13 shows the high temperature behavior for transverse mode, Cr/Au electroded 7/65/35and 8/70/30. Again, as in the case of the Cr/Au 9/65/35, the UV bias exposure acts to lower the light output and severely degrades the device.

A summary of the previous effects is illustrated in Figure 14. In case A at room temperature, bias voltage and UV exposure act to increase light output but retain some RLF at zero field. The net result of this effect, termed a residual memory effect, is to cause an apparent shift in the I vs. E loop to the opposite polarity, somewhat similar to the hysteresis loop shift for the FE 7/65/35 material in Figure 4. At high temperatures, the UV/bias exposure acts to lower the output in an apparent loop shift to the same polarity, again causing

RLF. Light output decreases significantly in the bias direction but is observed to increase noticeably in the opposite direction.

Visual confirmation of the high temperature (100°C) space charge related UV/bias effects are shown in Figure 15(A) for a Cr/Au transverse 9/65/35 (see also Figure 12), and Figure 15(B) for an ITO, transverse 9/65/35 (see also Figure 11). Considering case (A), the series of photographs from top to bottom are (1) a refreshed shutter in the on condition, (2) a 3 minute, UV/bias exposure condition on the shutter with the electrodes adjacent to the bright lines (photorefractive effect) biased positive during exposure and (3) bias field polarity is reversed and bright lines move to adjacent electrodes, thus still associating the bright lines with the positive electrodes.

In the case of (B), the series include (1) a refreshed shutter in the on condition, (2) a 3 minute UV/bias exposure condition on the shutter with the electrodes adjacent to the bright lines biased negatively during exposure and (3) bias field polarity is reversed and bright lines move to adjacent electrodes, thus still associating the bright lines with the negative electrodes. This behavior is exactly opposite to case (A), however, the reason for the different results is not yet known. In referring to Figure 11 for this same device, one can also see that it is truly remarkable in that its light output actually increases with bias voltage and UV exposure at high temperature.

As noted in the high temperature measurements, external field neutralization due to space charge generation, drift and trapping near an electrode can occur in relatively short periods of time; i.e., in a matter of seconds. Reversing the polarity of the electric field causes a similar behavior at the opposite electrode, leading one to believe that if polarity reversal was faster than the space charge drift, relatively insignificant movement of the space charge would take place during each cycle. Estimating the fastest time for space charge cancellation of approximately 5 seconds at 100°C, a 60 Hz reversal frequency was considered very adequate. Using this frequency, a series of tests were conducted on a transverse-mode, interdigitally electroded 9/65/35 sample. Effects of UV, temperature and ac vs. dc were evaluated in terms of on/off contrast ratio for times ranging from one minute to one week. Table 1 summarizes the results.

Table 1. Contrast Ratio Values of a 9/65/35 Shutter

ADDRESSING CONDITIONS		1 MINUTE		30 MINUTES		1 DAY		1 WEEK	
		25°C	100°C	25°C	100°C	25°C	100°C	25°C	100°C
UV LIGHT ON	DC	94	3	21	1	—	—	—	—
	AC	132	208	125	167	128	175	120	182
UV LIGHT OFF	DC	95	65	32	1	—	—	—	—
	AC	122	232	123	192	100	185	87	193

NOTE: ELECTRODES = .05 MM WIDE X .05 MM GAP
OPERATING VOLTAGE = 200V

$$CR = \frac{I_{ON}}{I_{OFF}}$$

From this test it can be concluded that (1) ac essentially eliminates the space charge problem at elevated temperatures, (2) UV is most detrimental at high temperatures in combination with a dc field, (3) contrast ratios are significantly better at elevated temperatures primarily because of a better off condition and (4) times up to one week have no noticeable degradation effect on shutter performance when ac operation was used. These results are in essential agreement with previously reported work by Samek and Munoz[29] who first noted the beneficial effects of ac operation for PLZT shutters.

Summarizing all of the dc shutter data in terms of contrast ratio as a function of temperature, Figure 16 clearly demonstrates that there is an optimum temperature range for best performance of a shutter; i.e., 40-60°C. Below this range, residual memory contributes to a degraded off condition, and above this range, the photoexcited space charge contributes to a poor on condition. Presumably, ac operation would help to correct the high temperature problem, but it would not materially affect the low temperature one.

VI. SUMMARY AND CONCLUSIONS

1. The PLZT ceramic materials are highly complex in nature and exhibit a wide variety of electrical, optical and UV radiation activated effects.

2. Residual memory effects in the non-memory PLZT materials are dominant at room temperature, whereas the photoconductivity and photoexcited space charge effects are significant at elevated (60-100°C) temperatures.

3. Photoexcitation studies at elevated temperatures seem to indicate that there are both negative (electrons) and positive (holes) carriers which are part of the total space charge and contribute to photoconduction.

4. Photoconduction in PLZT is most sensitive to wavelengths in the range of 360-380 nm.

5. A maximum in contrast ratio for shutter devices exists in the temperature range from 40-60°C

6. Many of the degradation effects in PLZT can be eliminated by means of ac operation.

REFERENCES

1. A. Sternberg, "Transparent Ferroelectric Ceramics: Properties and Applications," Ferroelectrics, <u>91</u> 53-67 (1989)

2. G.H. Haertling, "PLZT Electrooptic Materials and Applications - A Review," Ferroelectrics, <u>75</u> 25-55 (1987)

3. S. Brody, "Temperature Dependence of the Short Circuit Photocurrent in Ferroelectric Ceramics," Ferroelectrics, <u>10</u> 143-146 (1976)

4. K. Uchino, Y Miyazawa and S. Nomura, "High-Voltage Photovoltaic Effect in $PbTiO_3$-Based Ceramics,

Japanese J. Appl. Phys., 21 (12) 1671-1674 (1982)

5. A.E. Krumins and M. Vanecek, "Photoconductivity of Transparent Lanthanum - Modified Lead Zirconate Titanate Ferroelectric Ceramic," Phys. Stat. Sol. (a), 33 K31-34 (1976)

6. M.E. Lines and A.M. Glass, in Principles and Applications of Ferroelectrics and Related Materials, pp.453-466, Clarendon Press, Oxford (1977)

7. C.E. Land and P.S. Peercy, "A Review of the Effects of Ion Implantation on the Photoferroelectric Properties of PLZT Ceramics," Ferroelectrics, 45 25-43 (1982)

8. K. Tanaka, Y. Hamakawa, K. Wakino and M. Murata, "Photochromic Effect in Fe-Doped PLZT Ceramics, J. Am. Ceram. Soc., 59 (11/12) 465-469 (1976)

9. K. Uchino, M. Aizawa and S. Nomura, "Photostrictive Effect in (Pb,La)(Zr,Ti)O_3," Ferroelectrics, 64 194-208 (1985)

10. H. Adachi, T. Kawaguchi, K. Setsune, K. Ohji and K. Wasa, "Electro-Optic Effects of (Pb,La)(Zr,Ti)O_3 Thin Films Prepared by rf Planar Magnetron Sputtering," Appl. Phys. Lett., 42 (10) 867-868 (1983)

11. G.H. Haertling and C.E. Land, "Hot Pressed (Pb,La)(Zr,Ti)O_3 Ferroelectric Ceramics for Electrooptic Applications," J. Am. Ceram. Soc., 54 (1) 1-11 (1971)

12. K. Carl and K. Geisen. "Dielectric and Optical Properties of a Quasi-Ferroelectric PLZT Ceramic," Proc. IEEE, 61f (7) 967-974 (1973)

13. E.T. Keve and A.D. Annis, "Studies of Phases, Phase Transitions and Properties of Some PLZT Ceramics," Ferroelectrics, 5 77-89 (1973)

14. A.H. Meitzler and H.M. O'Bryan, Jr., "Polymorphism and Penferroelectricity in PLZT Ceramics," Proc. IEEE, 61 959-966 (1973)

15. G. Burns and F.H. Dacol, "The Observation of Glassy Polarization Behavior in Crystalline Ferroelectric Materials," Ferroelectrics, 52 103-113 (1983)ds

16. M. Yokosuka and M. Marutake, "Properties of Tetragonal PLZT with Reference to its Diffuse Phase Transition," Japanese J. Appl. Phys., 25 (7) 981-992 (1986)

17. P. Groves and P.A. Thomas, "Field-Enforced Behavior in 8.2/70/30 PLZT," Ferroelectrics, 89 179-192 (1989)

18. G. Smolensky, "Ferroelectrics with Diffuse Phase Transition," Ferroelectrics, 53 129-135 (1984)

19. L.E. Cross, "Relaxor Ferroelectrics," Ferroelectrics, 76 241-267 (1987)

20. C.E. Land, "Bistable Optical Information Storage Using Antiferroelectric Phase PLZT Ceramics," Advanced Ceramics, 6 (3) 10-15 (1988)

21. S. Mancha, J. Bullington, R. Carter and C. DeHainaut, "Antiferroelectric-Phase PLZT for Use in High Density Optical Data Storage," Ferroelectrics, 82 99-104 (1988)

22. F. Micheron, C. Mayeux, A. Hermosin and J.Nicolas, "Holographic Storage in Quadratic PLZT Ceramics," J. Am. Ceram. Soc., 57 (7) 306-308 (1974)

23. J.W. Burgess, R.J. Hurditch, C.J. Kirkby and G.E. Scrivener, "Holographic Storage and Photoconductivity in PLZT Ceramic Materials," Appl. Optics, 15 (6) 1550-1557 (1976)

24. F.S. Chen, "Optically Induced Change of Refractive Indices in $LiNbO_3$ and $LiTaO_3$," J. Appl. Phys., 40 3389 (1969)

25. A.E. Krumins and V.Y. Fritsberg, "Semiconductor Properties and General Applications of Optical Ceramics," Ferroelectrics, 35 149-154 (1981)

26. A.E. Krumins, "Electronic Processes and Phase Transitions in Transparent PLZT Ferroelectric Ceramics," Ferroelectric Letters, 1 89-95 (1983)

27. G.H. Haertling, "Photoferroelectric Effects in PLZT Ceramics," Paper No. T-17, Third US-Japan Seminar on Dielectric and Piezoelectric Ceramics, Toyama, Japan, November 9-12 (1986)

28. J.A. Bullington, M.D. Ivey and J.T. Evans, Jr., "Optical Characterization of Fatigue and Ageing in Ferroelectric Thin-film Capacitors," Proceedings of the First Symposium on Integrated Ferroelectrics, Colorado Springs, CO, March 28-30, 201-211 (1989)

29. N. Samek and M. Munoz, "Light Transmission Characteristics of a High Sensitivity Electrooptic Light Gate Array Based on PLZT Ceramics," Proc. SPIE, 37 (1981)

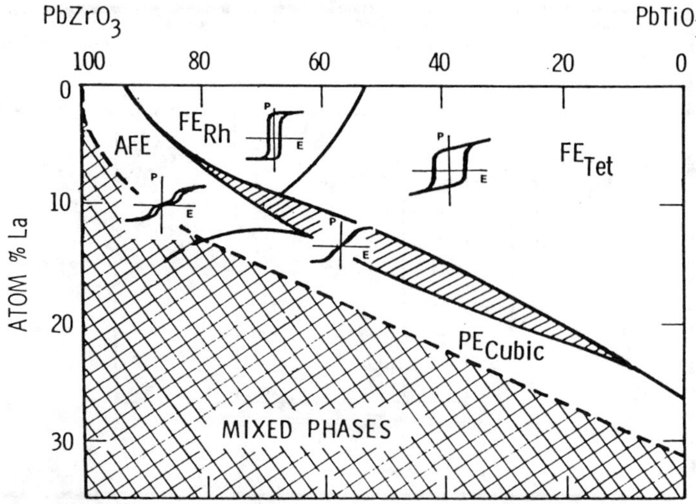

1. Room temperaure phase diagram of the PLZT system showing typical hysteresis loops for the various phases. The relaxor materials are located in the single cross-hatched region

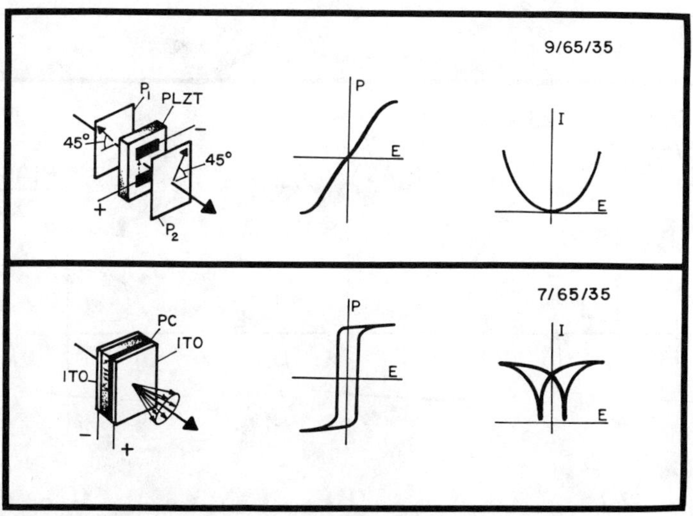

2. Electrooptic effects in PLZT ceramics illustrating operational configuraton, hysteresis loop and light intensity output for non-memory 9/65/35 (top) and memory 7/65/35 (bottom). The photoconductive (PC) layer on the 7/65/35 allows for spatially varying domain switching when storing images.

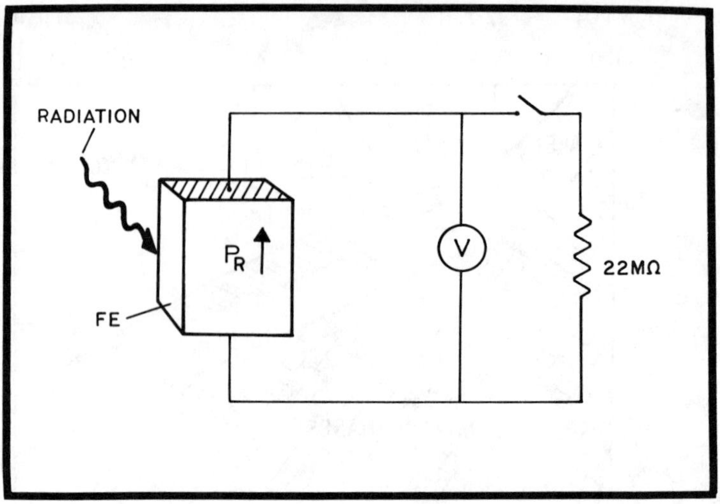

3. A typical setup for observing the photovoltaic effect in FE materials.

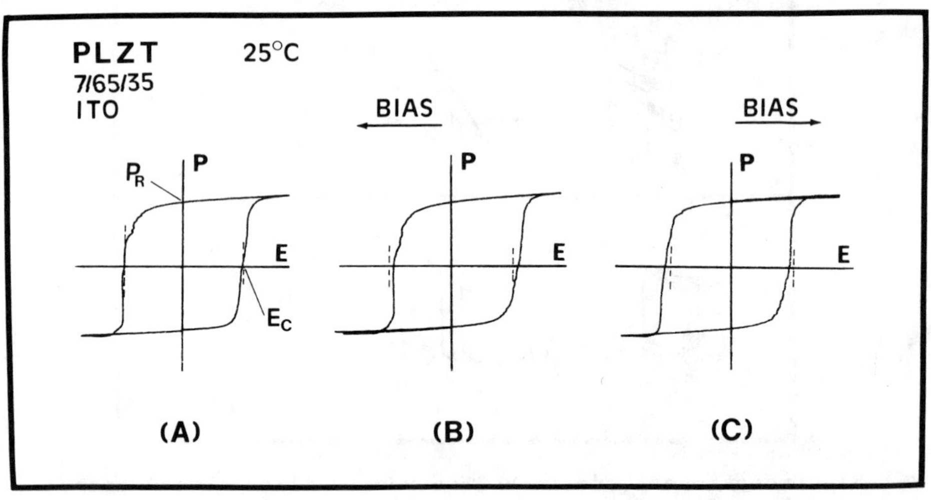

4. Hysteresis loop shift as a result of UV illumination. Note that the non-illuminated loop is symmetrical (A), which changes to an asymmetrical loop under UV illumination as in (B) and (C).

5. Example of a stored image in 7/65/35 utilizing the longitudinal scattering effect (cerampic). Courtesy of Sandia National Laboratories.

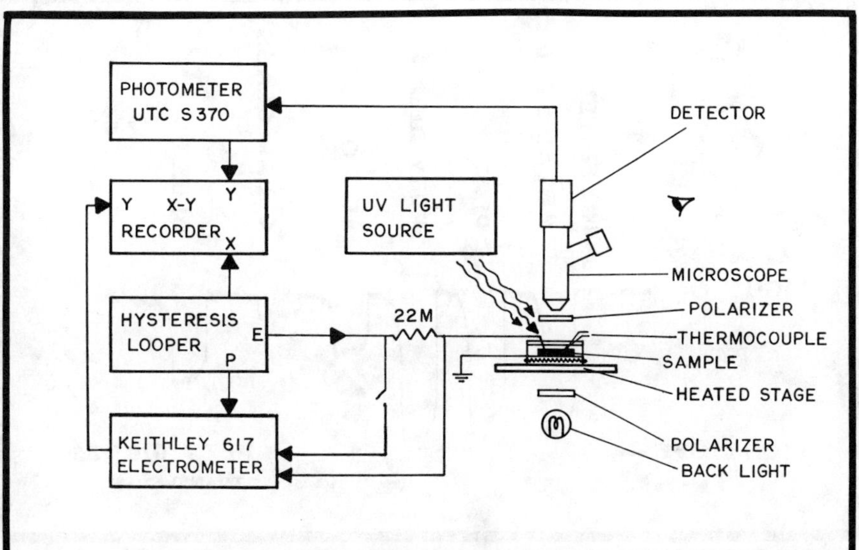

6. Setup for observing photoelectronic effects.

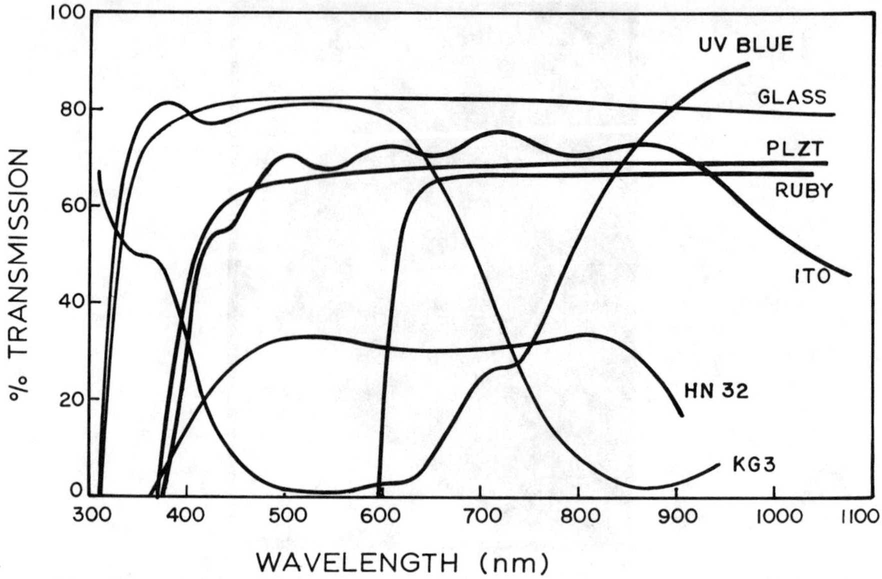

7. Spectral response of various filter materials

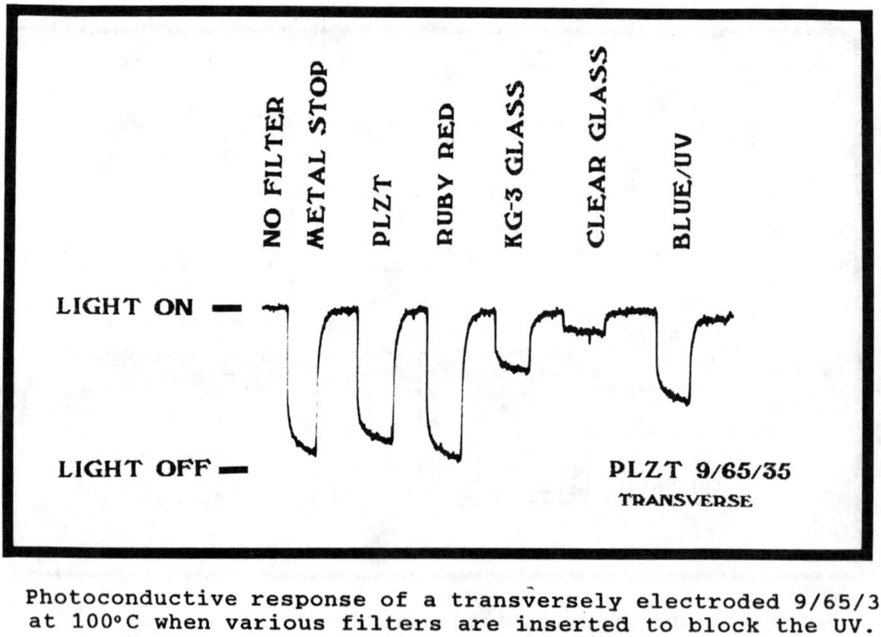

8. Photoconductive response of a transversely electroded 9/65/35 at 100°C when various filters are inserted to block the UV.

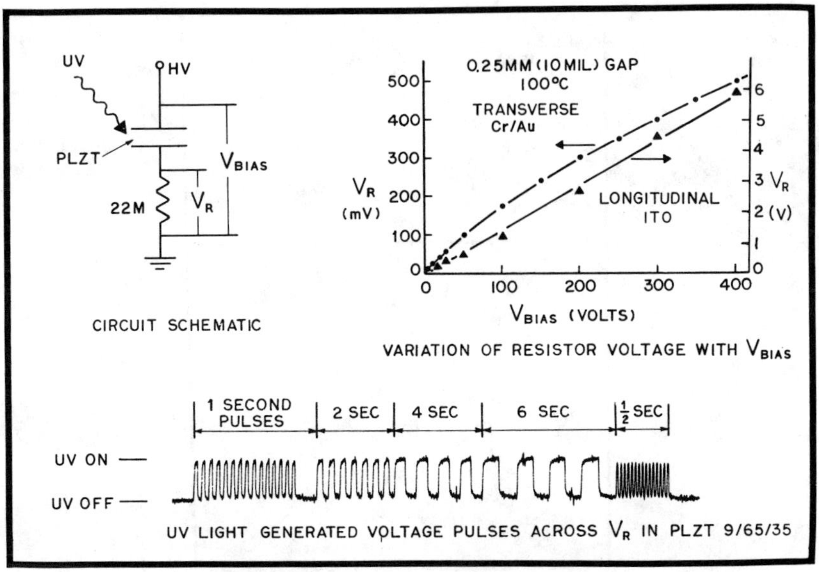

9. Photoconductivity effects in PLZT 9/65/35 showing the variation of these effects as a function of bias voltage (transverse and longitudinal) and response time.

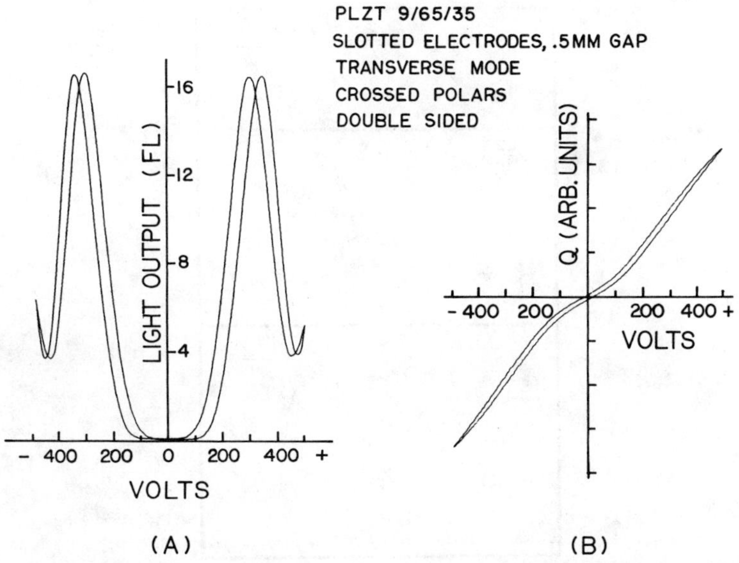

10. PLZT 9/65/35 shutter characteristics of light intensity output (A) and hysteresis loop (B).

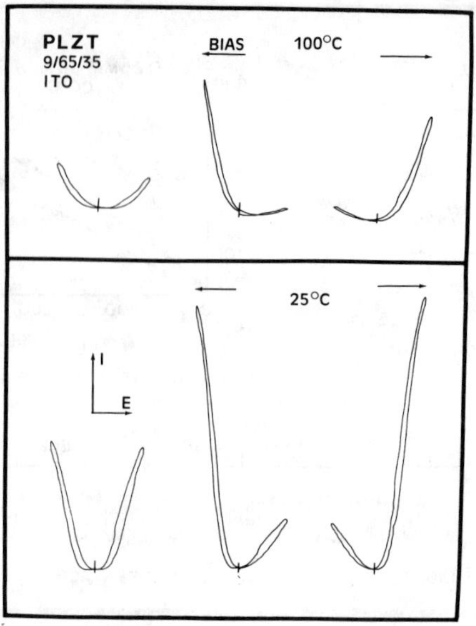

11. PLZT 9/65/35 shutter behavior for an ITO transversely electroded, unilluminated sample (left) and the same sample under UV illumination and electric field bias.

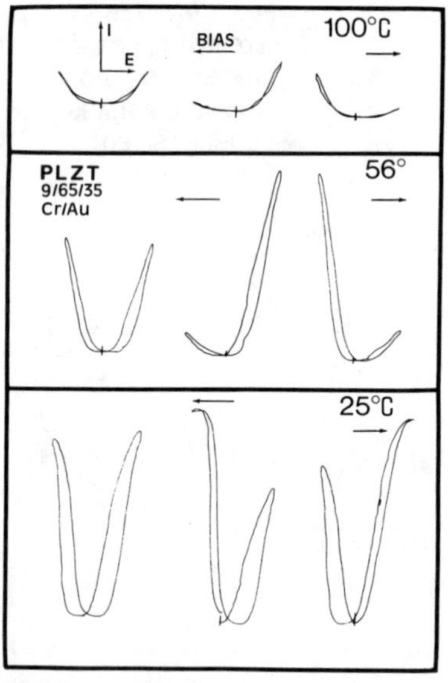

12. PLZT 9/65/35 shutter response as in Figure 11 but with Cr/Au electrodes.

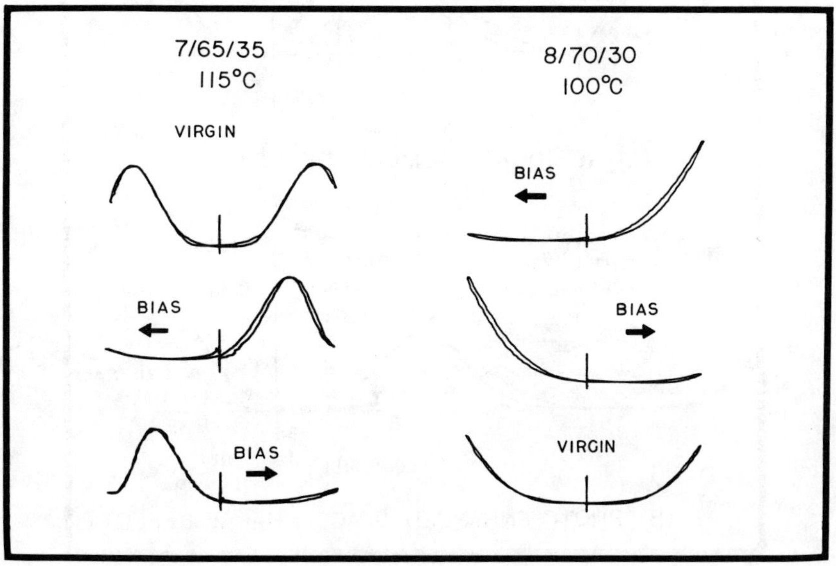

13. PLZT 7/65/35 and 8/70/30 shutter responses as a function of UV illumination and electric field bias at the indicated temperatures.

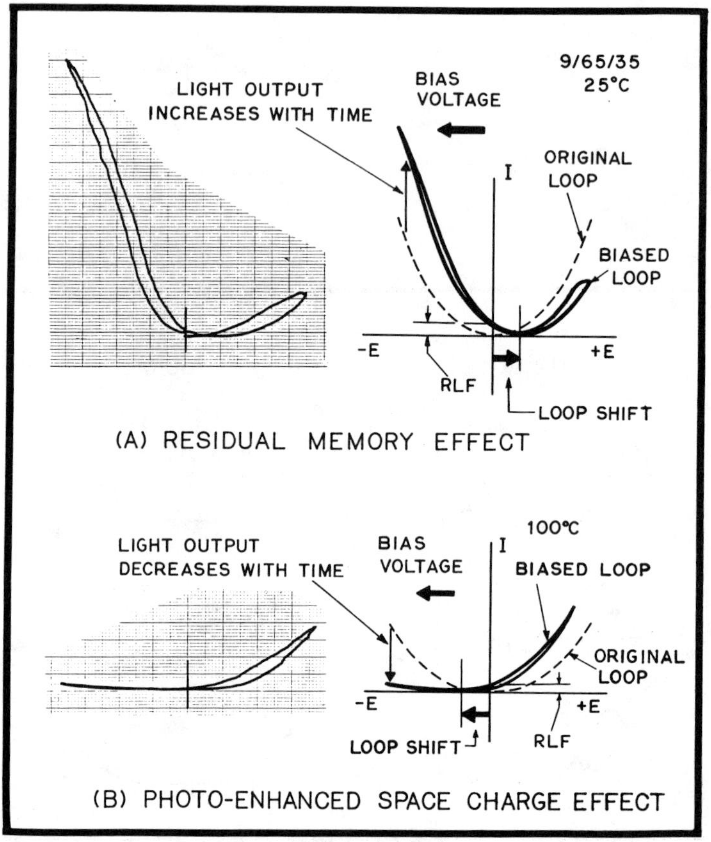

14. Models for RLF photoelectronic processes in PLZT non-memory ceramics, (A) photo-assisted residual memory effect and (B) photo-enhanced space charge effect.

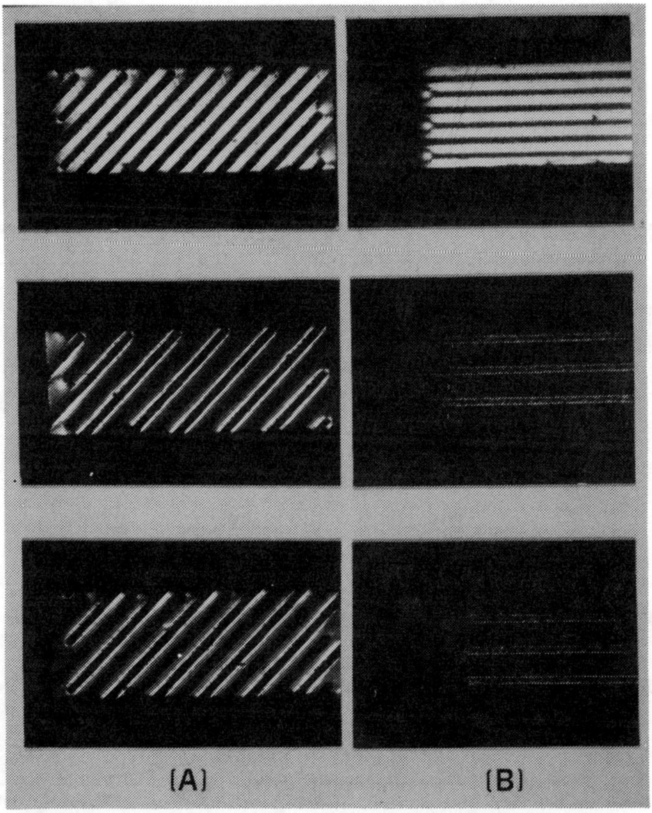

15. Photo-enhanced space charge effects in transversely electroded 9/65/35 shutters with Cr/Au (A) and ITO electrodes (B). Top photographs are for the refreshed shutter in the on condition and the bottom photographs show the change in the residual space charge condition (at zero bias and no illumination) after 3 minute UV/bias exposures of opposite polarity at 100°C.

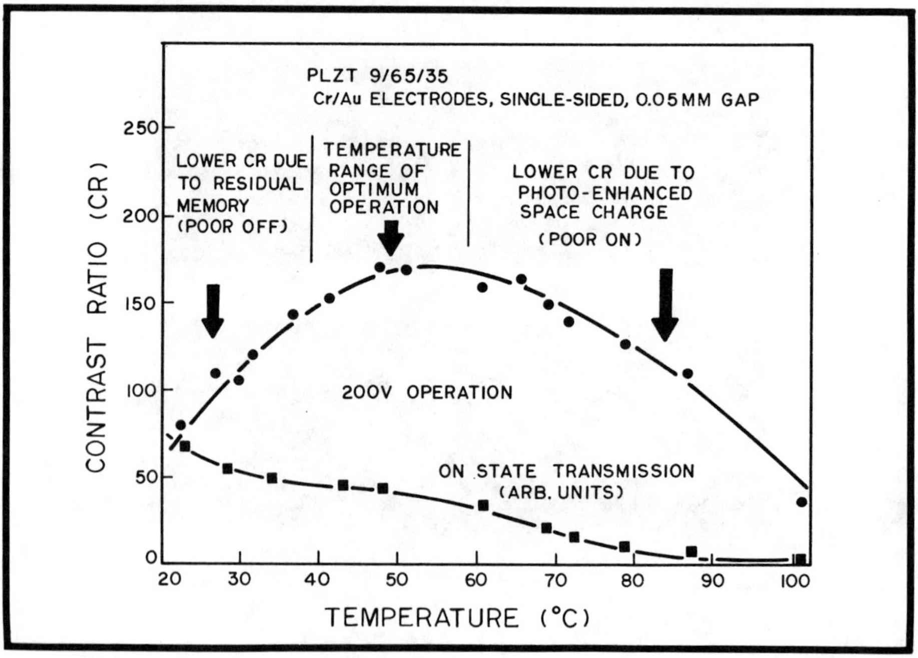

16. PLZT shutter characteristics as a function of temperature, illustrating that there is an optimum temperature range for continuously operating a shutter at its highest contrast ratio.

DEFECTS IN NONLINEAR OPTICAL OXIDE CRYSTALS

Patricia A. Morris
E. I. du Pont de Nemours & Co., Inc.
Central Research &Development Department
Experimental Station
Wilmington, DE 19880-0306

ABSTRACT

Defects in nonlinear optical oxide crystals can affect many of the properties required for device applications. The structures of typical crystals are nonstoichiometric on the cation sublattices (e.g. the A sublattice in crystals with the general formula AMO3). This may, at least in part, be due to the presence of the relatively strong covalent nature of the acentric oxide groups determining the nonlinear optical properties. These structures are also relatively tolerant with respect to occupancy on the cation sublattices. These circumstances make the incorporation of impurities into the lattice relatively easy and result in large distribution coefficients for many impurities during growth. Therefore, it is usually necessary to purify the starting materials of any unwanted ions. A combination of analytical techniques should be used to determine the impurities present in a material. For example, the effects of protons (OH^-) incorporated in the lattice of these crystals can be very detrimental and can be detected using infrared spectroscopy. The growth of many of these crystals requires flux techniques, but the temperature dependence of any nonstoichiometry present and the distribution coefficients may result in inhomogeneous distributions of defects and impurities in crystals grown using slow cooling techniques.

INTRODUCTION

Defects can affect many of the properties of single crystal oxides with large second order nonlinear optical susceptibilities ($\chi^{(2)}$). Applications of these types of materials include frequency convertors for laser systems, electro-optic modulators and switches, and holographic and phase conjugate optics.[1-5] The defects which may be present in nonlinear optical oxide crystals include 1) intrinsic atomistic defects (e.g. nonstoichiometry), 2) impurities, and 3) domains (many of these crystals are ferroelectric). The purpose of this paper is to point out defect problems related to the growth of nonlinear optical oxide crystals which impact the materials requirements for applications. Structural properties relating to intrinsic defect formation, impurity incorporation and crystal growth of several nonlinear optical oxides are discussed. Materials requirements, affected by intrinsic defects and impurities, for applications of crystals with large $\chi^{(2)}$ are examined and specific examples of defect and impurity effects on properties of $KTiOPO_4$, β-BaB_2O_4, $LiNbO_3$ and $BaTiO_3$ are described. Finally, areas requiring consideration for controlling the defects in nonlinear optical oxide crystals are summarized.

STRUCTURAL PROPERTIES

Induced Polarization and Origin of $\chi^{(2)}$

A polarization is induced in a material when subjected to laser radiation or dc electric fields. The following expression,[1-3]

$$P_i(\omega) = \varepsilon_o \left[\chi_{ij}^{(1)}(\omega) E_j(\omega) + 2\chi_{ijk}^{(2)}(\omega = \omega_1 + \omega_2) E_j(\omega_1) E_k(\omega_2) \right]$$

for the induced polarization (P_i) includes the first two terms in the series, where E is the electric field strength associated with the incident radiation or DC electric field, and ω is the frequency. The nonlinear optical coefficient ($d^{(2)}$) is defined by $1/2\,\chi^{(2)}$. The first term describes the linear optical effects: absorption, refraction, emission and reflection. The second term is responsible for the nonlinear polarization processes, such as second harmonic generation, parametric sum or difference mixing and the linear electro-optic or Pockels effect. Second harmonic generation and parametric generation are typically used to extend the frequency range of solid state lasers. The Pockels electro-optic effect is used in applications involving Q-switches for laser systems, optical modulators and switches, and the photorefractive effect for real-time holography and phase conjugation.

The basic structural units responsible for the second order nonlinear optical susceptibility in most oxide crystals are the acentric anionic groups.[4,6] The macroscopic $\chi^{(2)}$ is determined by the microscopic nonlinear susceptibility of the bonds in the acentric oxide group, the number and orientation of equivalent groups in a unit cell, and the number of unit cells per volume. The following acentric oxide groups contribute to $\chi^{(2)}$ in the oxide crystals listed: 1) MO_6, where typically M = Ti, Nb, Ta; $KTiOPO_4$, $BaTiO_3$, $LiNbO_3$, $KNbO_3$, $Sr_{1-x}Ba_xNb_2O_6$, $Ba_2NaNb_5O_{15}$, $K_3Li_2Nb_5O_{15}$, and $LiTaO_3$, 2) B_3O_6; β-BaB_2O_4 and B_3O_7; LiB_3O_5, 3) PO_4; KH_2PO_4 and $KTiOPO_4$, and 4) IO_3; $LiIO_3$.

Crystal Structures, Intrinsic Defects and Impurity Incorporation

Nonlinear optical oxide crystals with various crystal structures exist and their structures depend, in part, on the type of oxide group (e.g. MO_6, B_3O_6, PO_4) dominating the structure. The crystal structures of several of the nonlinear optical oxides are discussed below.

Many of the crystals containing MO_6 groups have either the perovskite (e.g. $BaTiO_3$), "perovskite-like", (e.g. $LiNbO_3$) or tungsten-bronze (e.g. $Sr_{1-x}Ba_xNb_2O_6$, $Ba_2NaNb_5O_{15}$, $K_3Li_2Nb_5O_{15}$) type structures.[4,7] The perovskite structure (AMO_3) consists of a set of MO_6 octahedra arranged in a simple cubic pattern, linked together by shared oxygen atoms.[7] The monovalent or divalent cation (A) occupies the spaces between them. The perovskite structure tolerates a degree of variation, especially with regard to the twelve coordinated site containing A ions.[8,9] The nonstoichiometry observed in $BaTiO_3$ occurs via vacancy formation on the Ba and O sublattices, as shown in Table I.[10,11,12] The "perovskite-like" structure (AMO_3) of $LiNbO_3$ and $LiTaO_3$ contains a sequence of distorted oxygen octahedra joined by their faces along a trigonal polar c-axis. Along the c-axis, the cations follow the sequence: (M, vacancy, A, M, vacancy, A ...).[7] The two cations present form anti-site defects in congruent nonstoichiometric $LiNbO_3$, which is Li-deficient (Table I).[13,14] The tungsten-bronze type structures (A_nMO_3 , where n = 0.5, 0.6 or 1) consist of a unit cell with 10 MO_6 octahedra linked by their corners in a way that forms three different types of tunnels running through the structure along the c-axis.[7] The MO_6 chains resemble those found in the perovskite structure. The A cations occupy the tunnels interstially in many ways depending on the composition. Three sites are present in the tunnels: twelve - coordinated A1 sites, nine - coordinated A2 sites, and three coordinated C sites. There are two different M sites. Three general compositional formulas exist. 1) $Sr_{1-x}Ba_xNb_2O_6$ contains Sr and Ba on the A1 and A2 sites with one of the six A1 and A2 sites and the C sites unfilled. Nonstoichiometry exists on the Ba and Sr sublattices over a range of x equal to 0.25 to 0.75.[15,16] The congruent composition is shown in Table I. 2) $Ba_2NaNb_5O_{15}$ has the A1 and A2 sites filled by Ba and Na, respectively, but the C sites are unfilled.[7] Nonstoichiometry exists in congruent crystals of $Ba_2NaNb_5O_{15}$ with an excess of Nb present, compensated by vacant Na sites (Table I).[17,18] 3) $K_3Li_2Nb_5O_{15}$ is expected to have all the sites filled, but this compound is only stable with an excess of Nb.[19] Nonstoichiometric crystals result and have vacant K and Li sites present (Table I).[7,19]

Table I. Intrinsic Nonstoichiometry and Defect Mechanism in Several Nonlinear Optical Oxide Crystals

Crystal	Nonstoichiometry*	Defects**	References
$BaTiO_3$	$Ba_{(1-x)}TiO_{(3-x)}$: $x \cong 0.01$	$V_{Ba}^{//}$, $V_O^{\bullet\bullet}$	[10,11,12]
$LiNbO_3$	$Li_{(1-x)}NbO_3$: $x = 0.036$	$Nb_{Li}^{\bullet\bullet\bullet\bullet}$, $V_{Nb}^{/////}$	[13,14]
$Sr_{(1-x)}Ba_xNb_2O_6$	$Sr_{(1-x)}Ba_xNb_2O_6$: $x = 0.39$		[15,16]
$Ba_2NaNb_5O_{15}$	$Ba_2Na_{0.72}Nb_5O_{15}$	$V_{Na}^{/}$	[17,18]
$K_3Li_2Nb_5O_{15}$	$K_{2.786}Li_{1.989}Nb_5O_{15}$	$V_K^{/}$, $V_{Li}^{/}$	[7,19]
$KTiOPO_4$	$K_{(1-x)}TiPO_{(5-x/2)}$: $x \cong 0.0005$	$V_K^{/}$, $V_O^{\bullet\bullet}$	[25]
β-Ba_2BO_4	?	?	

* Representative of the intrinsic nonstoichiometry present in as-grown crystals.
** Defects are presented as fully ionized.

The crystal structures of nonlinear optical oxides having the perovskite, "perovskite -like" and tungsten-bronze type structures have a relatively large degree of tolerance, especially with respect to the occupancy of the A cation sublattices.[7,8] Nonstoichiometry exists over the range of a few percent (e.g.$BaTiO_3$, $LiNbO_3$) or more (e.g.$Sr_{1-x}Ba_xNb_2O_6$).[10,13,15,17,19] The nonstoichiometry in these materials occurs at least in part due to the presence of the MO_6 groups, which are also primarily responsible for the second order nonlinear optical susceptibility. The relatively large contribution of covalent bonding in the MO_6 groups to the total lattice energy allows the structures to accommodate the nonstoichiometric defects occurring on the A cation sublattice with a relatively small cost of energy.[13] As shown in Table I, the nonlinear oxide crystals are typically nonstoichiometric with an excess of the M cation present. When a congruently melting composition exists, it is typically M-rich, showing that the free energy curve is asymmetric and skewed to lower energies toward the nonstoichiometric M-rich phase. The tolerance of these crystal structures and the presence of nonstoichiometry in the crystals allows the formation of solid solutions with a variety of ions or ion pairs. (Table II). Impurity substitution in the crystal lattice depends on the size, charge misfit, bonding and coordination preference of the impurity.[21] Isovalent impurity substitution is relatively easy for many ions due to the tolerance typically found in these structures.

Table II. Ions Forming Solid Solutions in Several Nonlinear Optical Oxide Crytals

Crystal	Solid Solution Forming Ions or Ion Pairs	References
$BaTiO_3$	Sr^{2+}, Pb^{2+}, (Na^{1+}+Nb^{5+}), (La^{3+}+In^{3+}), Ce^{3+}, Ca^{2+}, Si^{4+}, Zr^{4+}, Ge^{4+}, La^{3+}	[20]
$LiNbO_3$	Ta^{5+}, (Mg^{2+}+Ti^{4+})	[20]
	Mg^{2+}, Co^{2+}, Zn^{2+}, Cr^{3+}, Sc^{3+}, Sn^{4+}	[22]
$Ba_2NaNb_5O_{15}$	La^{3+}, Ti^{4+}, W^{6+}	[20]
	Li^{1+}, K^{1+}, Sr^{2+}, Ca^{2+}, Ta^{5+}	[23]
$KTiOPO_4$	Rb^{1+}, Tl^{1+}, NH_4^{1+}, As^{5+},	[27,28]
	Na^{1+}, Ag^{1+}, Cs^{1+}	[29]

[8,9] Aliovalent impurities are more readily incorporated in the lattice in the presence of charged nonstoichiometric defects, which can compensate the charge differences. For example, transition metals and rare earths can be incorporated during the growth of congruent $Li_{(0.964)}NbO_3$ crystals in amounts of up to several mole per cent and about 1 mole per cent, respectively.[22]

$KTiOPO_4$ contains both TiO_6 and PO_4 groups. The framework is characterized by TiO_6 octahedra in chains oriented along the [011] and [0$\bar{1}$1] directions, linked together by PO_4 bridges in an orthorhombic lattice.[24] The K ions are located in two sites relative to the two-fold screw axis, along the z-axis, and are surrounded by eight or nine oxygen ions, respectively. Two Ti and P sites are present in the lattice, as well. In the perfect crystal lattice all the K sites are filled, but experimental evidence indicates that intrinsic nonstoichiometry in $KTiOPO_4$ exists (Table I) to a very limited extent, on the K and O sublattices:[25]

$$KTiPO_5 \longrightarrow K_{(1-x)}TiPO_{(5-x/2)} + (x/2)K_2O$$

at levels corresponding to several hundred parts per million in crystals grown near the Curie temperature (936 °C [26]). This nonstoichiometry is observed to be very temperature dependent over the range of temperatures practical for flux growth.[25] Several isovalent ions form solid solutions with $KTiOPO_4$ (Table II), showing that the structure is relatively tolerant, with respect to isovalent impurities. But due to the relatively limited range of nonstoichiometry in the crystals, aliovalent impurities, such as divalent Ba, Sr and Ca, which substitute on the K site, are incorporated at concentrations less than one mole percent through ion exchange in nitrate melts.[30]

β-BaB_2O_4 (the low temperature form) contains B_3O_6 groups forming rings in sheets that alternate with Ba ions along the three-fold z-axis in stepped layers.[31,32] Two Ba sites, coordinated to seven oxygen ions, are present. Little is known about the defect structure and solubility of impurity ions in β-BaB_2O_4 crystals. If nonstoichiometry is present, it is most likely present with B excess due to the strong covalent nature of the B-O bonds. The absorption effects of neutral impurities contained in crystals have been studied, but the specific impurity ions present and their concentrations have not been reported.[33] Their presence indicates that this structure may also be relatively tolerant with respect to isovalent impurity incorporation.

CRYSTAL GROWTH

Nonlinear optical oxide crystals are grown by a variety of techniques. Nonstoichiometric crystals are grown using the Czochralski technique if congruently melting compositions are chosen (e.g. $Li_{(0.964)}NbO_3$, $Sr_{(0.61)}Ba_{(0.39)}Nb_2O_6$, $Ba_2Na_{(0.72)}Nb_5O_{15}$).[13,16,34,18] But many crystals require flux growth to obtain a low temperature phase (e.g. $BaTiO_3$ [35], β-BaB_2O_4 [36]) or prevent decomposition during melting (e.g. $KTiOPO_4$ [28]). The typical flux technique used involves either submerged or top-seeding and slow cooling of the solution. $KTiOPO_4$ is grown hydrothermally in addition to flux techniques.[28,37-43] The particular growth technique and conditions used have effects on defect formation and impurity incorporation during the growth process.

Nonstoichiometry

Nonstoichiometric crystal compositions grown by the Czochralski technique are chosen primarily because they are congruently melting. As previously discussed, the congruently melting compositions are M-rich (where the general compositional formula is A_nMO_3). To aid in maintaining the stoichiometry desired and obtaining a homogeneous crystal, low temperature gradients at the growth interface are typically used.[13,34] Variations in the nonstoichiometry in these crystals result from temperature fluctuations at the growth interface or insufficient mixing in the melt. Extremely stable growth conditions are necessary to produce crystals of optical quality.

The flux technique most often used involves either submerged or top-seeding and slow cooling of the solution. Therefore, crystals grown by this technique which are intrinsically nonstoichiometric will inherently contain inhomogeneous distributions of defects due to the growth occuring over a range of temperatures and the temperature dependent nature of any nonstoichiometry present. The inhomogeneous distribution of defects in crystals, which are intrinsically nonstoichiometric and grown by slow cooling techniques, is an important factor when considering the properties of these crystals.

Impurity Incorporation

The optimum conditions to obtain minimum incorporation of impurities, having equilibrium distribution coefficients less than one, during crystal growth include: 1) a slow growth rate, 2) maximum stirring and 3) a high liquid diffusivity.[44] To achieve a homogeneous distribution of impurities in a crystal, stable growth conditions and minimum temperature fluctuations are necessary. Generally, the growth conditions required for nonlinear optical oxide crystals are compatible with the conditions to achieve minimal incorporation of impurities during the growth process. However, some conditions exist in the flux growth of nonlinear optical oxides which can lead to impurity incorporation or inhomogeneous distributions of impurities in the crystals. Some flux compositions contain impurity ions (e.g. 15 mol percent Na_2O is used for flux growth of β-BaB_2O_4 [36]). Fluxes containing only the crystal constituents (e.g. $K_6P_4O_{13}$ flux is typically used for growth of $KTiOPO_4$ [28,40-43]) are preferred. Flux growth yields faceted interfaces which can have different distribution coefficients at different crystallographic planes and result in inhomogeneous concentrations of impurities in the crystals.[44] Slow cooling flux techniques can also lead to inhomogeneous concentrations of impurities due to temperature dependent defect structures and distribution coefficients. These conditions can clearly have effects on the distributions of intentional dopants, as well.

The major problem concerning impurity incorporation during the growth (Czochralski or slow cooling flux) of the typical nonlinear optical oxide crystals is that the distribution coefficients of impurity ions that can easily be accommodated in the lattice, are relatively high and these impurities are readily incorporated into the crystal during growth. For example, the following are distribution coefficients of several impurity ions in Czochralski grown $LiNbO_3$: Co^{2+} : 1.0, Fe^{3+} : 0.7, Rh^{3+} : 0.8, Si^{4+} : 0.5, Ti^{4+} : 0.7 and Sn^{4+} : 1.0. [22] As discussed above, the tolerant nature of the crystal structures and the presence of intrinsic nonstoichiometry allow relatively easy incorporation of many impurities into the lattice. Zone growth using a CO_2-laser for heating has been used to grow a variety of nonlinear optical oxide crystals with fiber morphologies [45,46] and could be used for zone refining of impurities. But the relatively high distribution coefficients of many impurities in nonstoichiometric nonlinear oxides means that the use of zone refining of these materials may be of limited value for purification purposes. These circumstances require that the starting materials for crystal growth be "pure" with respect to any unwanted impurities.

Domain Fomation

Single domain crystals can usually be obtained by stable growth below the Curie temperature using a single domain seed or by poling if growth occurs above the

Curie temperature. But periodic laminar domains are desired for some applications and can intentionally be produced in ferroelectric nonlinear optical oxide crystals. A one to one correspondence has been discovered between growth striations and domain walls formed in undoped and doped Czochralski grown $LiNbO_3$.[47,48] In Y-doped $LiNbO_3$ the domain walls form where the Y concentration gradient changes sign.[49,50] Periodic temperature variations have been used to form the concentration gradients in these crystals.[50] The impurity added should have a large distribution coefficient and the concentration fluctuations should be sufficiently high to form strong electric fields at the growing interface. To obtain strong electric fields the impurity ion should be aliovalent and there should be a pronounced difference in the diffusion coefficients of the impurities and compensating defects formed in the crystal.[50] Obviously, to avoid domain formation in crystals during growth, the absence of these conditions is desired.

EFFECTS ON MATERIAL REQUIREMENTS

The materials requirements of crystals to be used for applications involving large $\chi^{(2)}$ are dependent on the specific device and application.[1-5] The following property requirements can be affected by the defect and impurity contents in the crystal: 1) large $\chi^{(2)}$ or $d^{(2)}$, 2) optically transparent in the wavelength of interest, 3) low ionic and electrical conductivity for electro-optic and waveguide devices 4) high optical damage threshold in frequency generation and Q-switching applications, 5) high sensitivity and fast response time of the photorefractive effect for holographic and phase conjugate optics and 6) homogeneity with respect to the optical properties and conductivity.

Large $\chi^{(2)}$ or $d^{(2)}$

The following example shows the effects that some impurities can have on $\chi^{(2)}$ or $d^{(2)}$. Na and Ag are soluble in $KTiOPO_4$ via ion exchange processing to concentrations of 65 and 85 mole per cent, respectively.[29] The powder second harmonic intensities, relative to quartz, of $KTiOPO_4$ crystals containing several concentrations of these impurity ions are all lower than of undoped $KTiOPO_4$. For example, the second harmonic intensity of $Ag_{(0.5)}K_{(0.5)}TiOPO_4$ and $Ag_{(0.85)}K_{(0.15)}TiOPO_4$ are one and three orders of magnitude less than in $KTiOPO_4$. Refinement of the crystal structure of $Ag_{(0.85)}K_{(0.15)}TiOPO_4$ shows that no alteration in the Ti - O bond distances occurs to account for the large change in second harmonic intensity produced. The Ag ions are site selective with Ag filling 100% of the K(1) sites and 70% of the K(2) sites. The strong coordination of Ag(1) to the Ti chain oxygen results in a reduction of the Ti(1) - O - Ti(2) bond angle to 129.5^{o} from 135.5^{o}, found in $KTiOPO_4$. This perturbation of the Ti - O bond polarizability is believed to be responsible for the loss of second order nonlinearity. Thirty-two out of thirty-nine isostructural compositions formed in this study had lower second harmonic intensities than $KTiOPO_4$. The only ones which yielded comparable powder second harmonic intensities were $RbTiOPO_4$, $Rb_{(0.5)}K_{(0.5)}TiOPO_4$, $TlTiOPO_4$, $KTiOAsO_4$, $KTiO(PO_4)_{0.5}(AsO_4)_{0.5}$, $TlTiOAsO_4$ and $RbTiOAsO_4$.[51,52]

Recent work has shown that $KTiOAsO_4$ shows a significant enhancement of the nonlinear optic and electro-optic coefficients versus those of $KTiOPO_4$.[53] The effective nonlinear optic coefficient ($d_{eff}^{(2)}$) is 60% higher and the electro-optic coefficients are 10 - 30 % higher than those of $KTiOPO_4$. The structure of $KTiOAsO_4$ has a smaller acentricity of the TiO_6 octahedra and longer Ti - O bonds than in $KTiOPO_4$.[29] The structural differences in the TiO_6 octahedra do not explain the observed nonlinear optical enhancement, suggesting that the AsO_4 group makes a significant contribution to the nonlinear susceptibility [29] or that changes in the Ti-O bond nonlinearities occur [53].

While the results discussed here pertain to large concentrations of dopant ions to the $KTiOPO_4$ structure, it is indicative of the types of changes in $\chi^{(2)}$ or $d^{(2)}$ that can occur to a limited extent with the addition of impurities at lower concentrations. As seen above [29], most dopants in the $KTiOPO_4$ structure (other than Rb, Tl, and As) reduce the observed second harmonic intensities.

Optical Transparency in the Wavelenth Range of Interest

Optical transparency of nonlinear optical oxide crystals in the wavelength range of interest is critical for use of the material. This example shows the limitations that impurity absorption can impose on the range of device applications.

β-Ba_2O_4 is a harmonic generation material used in the ultra-violet (UV) wavelength region.[4] The UV intrinsic absorption edge at room temperature is 190 nm.[33] To use this material with high efficiency in the UV range the transmittance from 200 to 350 nm should be high. The temperature dependence of the UV absorption spectrum of a typical β-Ba_2O_4 crystal from 300 to 78 K shows that three absorption steps appear at about 10 ^oC and increase rapidly up to saturation with decreasing temperature.[33] The transmittance of a 1 cm thick crystal at 222 nm falls below 22% at 0 ^oC. This restricts the temperature range of use in the UV of β-Ba_2O_4 for practical applications. The low-temperature absorption steps are attributed to the presence of a donor energy state induced by certain impurity centers in the crystal, but the specific impurities responsible were not reported.

Low Ionic and Electrical Conductivity

High ionic or electrical conductivity can be very harmful for a variety of electro-optic devices. Also, if a high ionic conductivity is present in any waveguide device material, the waveguides may not be stable during device processing or in an electric field.

Control of the ionic conductivity of $KTiOPO_4$ is important in both the processing of waveguides and to electro-optic waveguide device stability.[54] A moderate ionic conductivity is necessary to form waveguides in the material, but if excessive the mode distribution in the waveguides can be altered during device processing.[54,55] In $KTiOPO_4$, one-dimensional K ion conductivity occurs along the two-fold screw axis.[25,26,56] The ionic conductivity of $KTiOPO_4$ is found to vary by four to five orders of magnitude, depending on the crystal growth technique (flux, high- and low- temperature hydrothermal) and conditions of growth.[25] Impurity analyses of typical impurities in $KTiOPO_4$ crystals, grown by both flux and hydrothermal techniques, reveal no correlation to the ionic conductivity of the material. The variation in ionic conductivity of flux crystals grown over different temperature ranges is due to nonstoichiometric defects present on the K and O sublattices, described above. But the concentrations of vacant potassium sites formed as a result of the nonstoichiometry at the temperatures involved in hydrothermal growth are not sufficient to explain the ionic conductivity of these crystals. Infrared spectra show a correlation of higher AC conductivity to the distribution of OH^- sites in hydrothermally grown crystals. OH^- incorporation in $KTiOPO_4$ can compensate for the formation of vacant K sites and increase the ionic conductivity. Further work is being done to determine the structural location of the specific OH^- sites which contribute to the ionic conductivity in hydrothermal $KTiOPO_4$.

An important aspect of this example is that standard analyses of typical impurities in a material are not necessarily sufficient to determine all the impurities present which may be affecting its properties. Proton impurities in nonlinear optical oxide crystals were previously shown to produce inhomogeneous refractive indices in $LiNbO_3$ [57,58] and play a role in polarization reversal in $Ba_2NaNb_5O_{15}$ [59], Protons have now been found to increase the ionic conductivity in $KTiOPO_4$.[25] These results indicate the importance of considering proton impurities in addition to typical impurities, detected by standard analytical techniques, when considering the effects that impurities have on a material's properties.

High Damage Threshold

Optical damage in nonlinear optical oxide crystals is typically photorefractive (e.g. $LiNbO_3$ [60], $BaTiO_3$ [61]) or photochromic (e.g. $KTiOPO_4$ [62,63]) and limits the useable laser power levels in frequency generation and Q-switching device applications. The photorefractive process, discussed in the following section, has been widely studied [64] and defects or impurities shown to be directly responsible for the photorefractivity of a variety of materials (e.g. Fe impurities in $LiNbO_3$ [60]).The optical damage threshold of a material, that damages via the photorefractive process, can be increased by eliminating the defects or impurities responsible.

Optical damage thresholds for $KTiOPO_4$ are reported to be 1 - 30 GW/cm^2.[54] Photochromic damage is observed as the creation of "gray tracks", which are dichroic and preferentially absorb both the fundamental and second harmonic generated light with their electric vector along the Z-direction.[63,65] Damage with similar optical absorption properties can be created by electric fields or hydrogen annealing, which is attributed to Ti^{3+} formation in the crystal.[63,66] A correlation is observed between electric field induced damage and ionic conductivity indicating that the defects responsible for the ionic conductivity in $KTiOPO_4$ also contribute to the damage process.[25] Further work is being conducted to determine the precise mechanism of damage in $KTiOPO_4$ crystals grown by flux and hydrothermal techniques.

High Sensitivity and Fast Response Time of the Photorefractive Effect

The photorefractive effect is used for a variety of applications, including holography and phase conjugation.[63] The photorefractive process involves: 1) the optical generation of carriers from donor or acceptor levels in the band gap, 2) the photoconductivity of the material to redistribute the excited carriers to areas which are less illuminated, where they are trapped and form space charge fields, and 3) the electro-optic coefficient of the material which produces the index of refraction change in the material resulting from the space charge fields present.[61] Photorefractivity has been observed in a variety of nonlinear optical oxide materials and several have the desired sensitivity (10^{-3} cm^2/J), but response times are generally slower than that required ($\leq$ 1 msec) for many applications.[67] The host crystal properties which affect the photorefractive characteristics of a material are the electro-optic coefficient, permitivity, dielectric relaxation, and structural flexibility.[67] The photoionizability, site preference and valence state of the dopant ion determines the carriers generated in the material.[67] A significant amount of work is currently being done to understand the mechanism of photorefractivity so that either the properties can be improved or other materials can be developed.[64]

Photorefractivity in $BaTiO_3$ has been widely studied due to its high electro-optic coefficient,[61,68,69] but the specific mechanism of photorefractivity in $BaTiO_3$ is still not clear. Fe impurities were originally considered to be responsible for the photorefractive response of $BaTiO_3$.[61] Recent work has shown that Fe impurities are not responsible for the photorefractivity in undoped $BaTiO_3$.[69] Examination of the photorefractive response of undoped $BaTiO_3$ crystals grown using several starting materials indicates that undetected impurities in the crystal are most likely the source of the photorefractivity.[68] Whether the specific absorbing mechanism involves a variable valence ion or an impurity-vacant oxygen complex is not known. The defect complex, $(Ni_{Ti}V_O)^{/}$, is consistent with the absorption spectrum.[68]

Evident from this example is the intimate relationship of the photorefractive effect to the defects or impurity concentrations in the crystal. To optimize the photorefractive sensitivity and response time the appropriate dopant or defect, with regard to photoionizabilty, site preference and valence state must be obtained in the crystal.[67] The materials must also not contain any defects or impurities which would adversely affect the photorefractive process.

Homogeneity with Respect to Optical Properties and Conductivity

Homogeneity of the properties (nonlinear optical susceptibility, optical absorption, index of refraction, optical damage susceptibility, photorefractivity, ionic and electrical conductivity) of bulk nonlinear optical oxide crystals is crucial for virtually all applications of these materials.[1-5] From the above examples, it is clear that the defect and impurity concentrations in the material introduced unintentionally or intentionally must be uniform throughout the material. Another important example is in the formation of waveguides which are necessary for a variety of electro-optic devices. Typically, an ion is diffused or exchanged into the lattice to produce an area of higher index and provide a guiding index of refraction structure. The control of the waveguide depth or shape is critical in determining the modal distribution in the waveguide and the performance of the optical device. Any inhomogeneities in the defects or impurities present which produce nonuniform bulk indices of refraction, diffusion coefficients or exchange rates of the ion to be added can affect the waveguide structure and resulting device performance. Therefore, crystal growth parameters associated with non-uniform distributions of defects or impurities (unstable growth rates, faceted interfaces and slow cooling flux growth techniques) should obviously be avoided if possible to assure maximum homogeneity of the crystal.

The requirement of uniformity has the important exception related to the formation of periodic laminar domains in ferroelectric nonlinear optical oxide crystals, discussed above. These domain structures are used to enhance second harmonic conversion efficiencies [70] and extend phase matching capabilities [71]. In these crystals, the periodic domain structures must be formed in a controlled manner with spacings equal to an odd multiple of the coherence length for second harmonic generation, typically on the order of a few microns. The issues pertaining to homogeneity, discussed above, are equally applicable to these materials, with the exception of the intentionally introduced concentration variation of the specific impurity responsible for the formation of the domains.

SUMMARY

As seen in this discussion, defects in nonlinear optical oxide crystals can affect many of the properties required for device applications. The structures of typical crystals are nonstoichiometric on the cation sublattices (e.g. the A sublattice in crystals with the general formula AMO_3). This may, at least in part, be due to the presence of the relatively strong covalent nature of the acentric oxide groups determining the nonlinear optical properties. These structures are also relatively tolerant with respect to occupancy on the cation sublattices. These circumstances make the incorporation of impurities into the lattice relatively easy and result in large distribution coefficients for many impurities during growth. Therefore, it is usually necessary to purify the starting materials of any unwanted ions. A combination of analytical techniques should be used to determine the impurities present in a material. For example, the effects of protons (OH^-) incorporated in the lattice of these crystals can be very detrimental and can be detected using infrared spectroscopy. The growth of many of these crystals requires flux techniques, but the temperature dependence of any nonstoichiometry present and the distribution coefficients may result in inhomogeneous distributions of defects and impurities in crystals grown using slow cooling techniques. These growth techniques are generally not recommended when uniformity of properties is required.

ACKNOWLEDGEMENTS

I would like to acknowledge John D. Bierlein and Thurman E. Gier for helpful discussions during the preparation of this manuscript.

REFERENCES

1 Y. R. Shen, The Principles of Nonlinear Optics (John Wiley & Sons, New York, 1984).
2 A. Yariv and P. Yeh, Optical Waves in Crystals (John Wiley & Sons, New York, 1984).
3 S. Singh, in: Handbook of Laser Science and Technology, Vol. III, Ed. M.J. Weber (CRC Press, Boca Raton, Florida, 1986).
4 C. Chen and G. Liu, Ann. Rev. Mater. Sci. 16 (1986) 203.
5 A. A. Ballman, R.L. Byer, E. Eimerl, R.S. Feigelson, B.J. Feldman, L.S. Goldberg, N. Menyuk and C.L. Tang, Appl. Optics 26 (1987) 224.
6 M. DiDomenco Jr. and S.H. Wemple, J. Appl. Phys. 40 (1969) 720, 735.
7 M. E. Lines and A. M. Glass, Principles and Applications of Ferroelectrics and Related Materials (Clarendon Press, Oxford, 1977).
8 J. B. Goodenough and J. A. Kafalas, J. Solid State Chem. 6 (1973) 493.
9 F. D. Bloss, Crystallography and Crystal Chemistry (Holt, Rinehart and Winston, New York, 1971).
10 D. E. Rase and R. Roy, J. Am. Cer. Soc. 38 (1955) 102.
11 N. G. Eror and D. M. Smyth, J. Solid State Chem. 24 (1978) 235.
12 J. F. Baumard and P. Abelard, Solid State Ionics 12 (1984) 47.
13 J. R. Carruthers, G. E. Peterson, M. Grasso and P.M. Bridenbaugh, J. Appl. Phys. 42 (1971) 1846.
14 S. C. Abrahams and P. Marsh, Acta Cryst., B42 (1986) 61.
15 J. R. Carruthers and M. Grasso, J. Electrochem. Soc. 117 (1970) 1426.
16 K. Megumi, N. Nagatsuma, K. Kashiwada and Y. Furuhata, J. Mater. Sci. 11 (1976) 1583.
17 J. R. Carruthers and M. Grasso, Mat. Res. Bull. 4 (1969) 413.
18 K. G. Barraclough, I. R. Harris, B. Cockayne, J.G. Plant, and A. W. Vere, J. of Mater. Sci. 5 (1970) 389.

19 S. C. Abrahams, P. B. Jamieson, J. L. Bernstein, J. Chem. Phys. 54 (1971) 2355.
20 Phase Diagrams for Ceramists, E. M. Levin, C. R. Robbins and H. F. McMurdie, Ed. M. K. Reser (Am. Cer. Soc., Columbus, OH, 1964).
21 K. Nassau, in: Ferroelectricity, Ed. E.F. Weller (Elsevier, Amsterdam, 1967).
22 A. Rauber, in: Current Topics in Mater. Sci., Vol. I, Ed. E. Kaldis (North Holland, Amsterdam, 1978).
23 L. G. Van Uitert, J. J. Rubin and W. A. Bonner, J. Quantum Elec. 4 (1968) 622.
24 I. Torjman, R. Masse and J. C. Guitel, Z. Kristallogr. 139 (1974) 103.
25 P. A. Morris, M. K. Crawford, A. Ferretti, R. H. French, M. G. Roelofs, J. D. Bierlein, J. B. Brown, G. M. Loiacono, G. Gashurov, in: Optical Materials: Processing and Science, Vol 152, Ed. D. B. Poker and C. Ortiz (Mater. Res. Soc., Pittsburg, PA, 1989).
26 V. K. Yanovksi and V. I. Voronkova, Sov. Phys. Solid State 27 (1985) 1308.
27 R. Masse and J. C. Grenier, Bull. Soc. Fr. Min. Cristallogr. 94 (1971) 437.
28 T. E. Gier, U. S. Patent No. 4,231,838 (4 November 1980).
29 G. D. Stucky, M. L. Phillips and T. E. Gier, submitted.
30 M. G. Roelofs and J. D. Bierlein, submitted to J. Appl. Phys.
31 L. S. Fang, H. M. Yun and H. J. Ling, Acta Physica Sinica 31 (1982) 948.
32 J. Leibertz and S. Stahr, Z. Kristallogr. 165 (1983) 91.
33 G. Zhang, Y. Yang and C. Zhang, Appl. Phys. Lett. 53 (1988) 1019.
34 R. R. Neurgaonkar, M.H. Kalisher, T. C. Lim, E. J. Staples and K. L. Keester, Mat. Res. Bull. 15 (1980) 1235.
35 V. Belruss, J. Kalnajs, A. Linz and R. C. Folweiler, Mat. Res. Bull. 6 (1971) 899.
36 A. Jiang, F. Cheng, Q. Lin, Z. Cheng and Y. Zheng, J. Crys. Growth 79 (1986) 963.
37 T. E. Gier, U. S. Patent No. 4,305,778 (15 December 1981).
38 R. A. Laudise, R. J. Cava and A.J. Coporaso, J. Crys. Growth 74 (1986) 275.
39 R. F. Belt, G. Gashurov and R. A. Laudise, SPIE Proc. 968 (1988) 100.
40 J. C. Jacco, G. M. Loiacono, M. Jaso, G. Mizell and B. Greenberg, J. Crys. Growth 70 (1984) 484.
41 P. F. Bordui, J. C. Jacco and G. M. Loiacono, J. Crys. Growth 84 (1987) 403.
42 S. Dezhong and H. Chaoen, Prog. Crys. Growth and Charact. 11 (1985) 269.
43 C. Defan and Y. Zhegtang, J. Crys. Growth 79 (1986) 974.
44 M. C. Flemings, Solidification Processing (McGraw-Hill, New York, 1974).
45 R. S. Feigelson, W. L. Kway and R. K. Route, Optical Eng. 24 (1985) 1102.
46 M. M. Fejer, J. L. Nightingale, G. A. Magel and R. L. Byer, Rev. Sci. Instrum. 55 (1984) 1791.

47 K. Nassau, H. J. Levinstein and G. M. Loiacono, J. Phys. Chem. Solids 27 (1966) 983.
48 J. C. Peuzin and M. Tasson, Phys. Stat. Sol. A 37 (1976) 119.
49 N. B. Ming, J. F. Hong and D. Feng, J. Mat. Sci. 17 (1982) 1663.
50 J. Chen, Q. Zhou, J. F. Hong, W. S. Wang, N. B. Ming and D. Fang, J. Appl. Phys. 66 (1989) 336.
51 J. D. Bierlein and T. E. Gier, U. S. Patent No. 3,949,323 (6 April 1976).
52 F. C. Zumsted, J. D. Bierlein and T. E. Gier, J. Appl. Phys. 47 (1976) 4980.
53 J. D. Bierlein, H. Vanherzeele and A. A. Ballman, Appl. Phys. Lett. 54 (1989) 783.
54 J. D. Bierlein, SPIE Proc. 994 (1988) 160.
55 V. Guerra and M. Chouinard (personal communication).
56 V. A. Kalesinkas and N. I. Pavlova, I. S. Reez and J. P. Grigas, Sov. Phys. Collect. 22 (1982) 68.
57 R. G. Smith, D. B. Fraser, R. T. Denton and T. C. Rich, J. Appl. Phys. 39 (1968) 4600.
58 J. Jackel, C. Rice and J. Veselka, Appl. Phys. Lett. 41 (1982) 607.
59 S. Singh, H. J. Levinstein and L. Van Uitert, Appl. Phys. Lett. (1970) 176.
60 H. Kurtz, E. Kratzig, W. Keune, H. Engelmann, U. Gonser, B. Dischler and A. Rauber, Appl. Phys. 12 (1977) 355.
61 M. B. Klein and R. N. Schwartz, J. Opt. Soc. Am. B 3 (1986) 293.
62 V. V. Lemeshko, V. V. Obukhovskii, A. V. Stoyanov, N. I. Pavlova, A. I. Pisanskii, and P. A. Korotkov, Ukr. Fiz. Zh. Russ. Ed. 31 (1986) 1745,
63 M. G. Roelofs, J. Appl. Phys. 65 (1989) 4976.
64 Photorefractive Materials, Effects, and Devices, Ed. G. C. Valley and P. Yeh, J. Opt. Soc. Am. B 5 (1988) 1685.
65 T. A. Driscoll, H. J. Hoffman, R.E. Stone, J. Opt. Soc. Am. B 3 (1986) 683.
66 V. V. Lemeshko, V.V. Obukhuovskii, A. V. Stoyanov, N. I. Pavlova, A. I. Pinsanskii, P. A. Korotkov, Ukr. Fiz. Zh. Russ. Ed., 31 (1986) 1745.
67 R. R. Neurgaonkar and W. K. Cory, J. Opt. Soc. Am. B 3 (1986) 274.
68 P. G. Schunemann, T. M. Pollak, Y. Yang, Y. Y. Teng and C. Wong, J. Opt. Soc. Am. B 5 (1988) 1702.
69 P. G. Schunemann, D. A. Temple, R. S. Hathcock, H. L. Tuller, H. P. Jenssen, D. R. Gabbe, and C. Warde, J. Opt. Soc. Am. B 5 (1988) 1685.
70 D. Feng, N. B. Ming, J. G. Hong, Y. S. Yang, J. S. Zhu, Z. Yang and Y. N. Wang, Appl. Phys. Lett. 37 (1980) 607.
71 R. L. Byer, Laser Focus World 3 (1989) 77.

NEW MATERIALS FOR HIGH AVERAGE POWER ELECTRO-OPTIC SWITCHES

Chris A. Ebbers and David Eimerl
Lawrence Livermore National Laboratory
P.O. Box 808, L-250
Livermore, Ca 94550

ABSTRACT

We will discuss the strengths and limitations of KDP and $LiNbO_3$ as electro-optic switch materials, as well as the physical properties a new electro-optic switch candidate must have in order to be useful. One area that we have examined to find new materials which meet some or all of those requirements will be described. We will also describe a new method we have developed for measuring the electro-optic coefficients of small crystalline samples 2 mm in size, and we will present data on new electro-optic materials taken using this new technique.

INTRODUCTION

As the average power of lasers continues to increase, we must reevaluate the materials used in the optical elements associated with them. Elements such as frequency doublers and electro-optic (EO) switches are routinely used in high average power (HAP) lasers. When design limitations occur due to constraints on material choices, one has the following options: Improve existing EO materials by increasing the damage threshold, reducing the photorefractive effect or eliminating impurity absorptions; or search for new materials which are inherently less prone to the problems of existing crystals. Both choices are viable options. We have chosen to search for new EO materials which will extend the capability of the next generation of HAP lasers.

Section I gives a brief introduction to the electro-optic effect. Section II describes the material requirements a HAP EO switch material must possess. In Section III we describe where we intend to search for new materials and in the Section IV we discuss the characterization of any new materials found.

I. THE ELECTRO-OPTIC EFFECT

The linear electro-optic effect is a linear change in the index of refraction of a material with respect to an applied electric field.[1] The linear susceptibility of a material is most conveniently represented by the index ellipsoid. This surface represents the index of refraction observed given a propagation direction and direction of incident polarization. For a cubic material and light of single frequency, this surface is a sphere, that is the index of refraction is a constant given any propagation/polarization direction. The polarization state of a light beam passing through a cubic material remains unchanged. If our cubic material also lacks a center of symmetry, when an electric field is applied the index surface will be distorted and the material will exhibit different indices of refraction along the preferred vibration directions. A light beam passing through this material is now broken into two rays, each travelling at a slightly different velocity. As the rays exit the crystal, they recombine and interfere with each other, resulting in a rotation of the original polarization state. The voltage required to rotate the polarization state by 90 degrees is known as the half-wave voltage:

$$V_\pi = \frac{\lambda\, l_{field}}{R\, l_{prop}} \qquad R \equiv n_o^3\, r_{xyz} \tag{1}$$

In general, the half wave voltage is proportional to the distance across which the field is applied and inversely proportional to the propagation distance and the generalized electro-optic coefficient, R. By increasing the propagation distance through the crystal, or by increasing the electro-optic coefficient, the half-wave voltage is reduced. Table I lists the propagation and field directions as well as the effective EO coefficient of various well characterized materials.

Table I. Propagation directions, field directions and effective EO coefficients of several well characterized materials

Material	$\vec{K}$	$\vec{E}$	R	Value(pm/V)
$LiNbO_3$	z	x	$R = 2n_o^3 r_{xxx}$	68.8
KD^*P	[110]	z	$R = n_o^3 r^s_{xyz}$	79.4
BBO	z	y	$R = 2n_o^3 r_{yyy}$	19.0
SiO_2	y	x	$R = n_o^3 r^s_{xxx}$	0.82

II. REQUIREMENTS OF A HAP EO SWITCH MATERIAL

The medium power solid state laser group at LLNL has produced a laser capable of generating 100 watts of average power at 532 nm with a peak power of 2 gigawatts. Any optical element, such as an EO switch, within that cavity must be able to sustain, without damage, the average as well as the peak power. An EO switch material used in such a device must have the following characteristics:

(1) A high damage threshold.

(2) Low optical absorption at the fundamental frequency.

(3) Low half-wave voltage.

(4) Good thermal and mechanical properties.

The damage threshold of the of the least durable element in the laser cavity will determine the beam aperture, since this places a limit on the peak power a material can withstand without damaging. The optical absorption in combination with the heat conductivity and fracture strength limit the maximum average power a material can withstand before shattering.[2] The half-wave voltage will determine the size and weight of the power supply used to drive the switch. Given a fixed power supply design and input aperture, this directly translates into demanding a piece of switch material of the proper length as determined by the magnitude of its effective EO coefficient.

Energy is deposited in the bulk of the switch in the form of heat. The dominant contribution to this heat depends upon the application. In telecommunication EO modulators, it is the fraction of modulation power absorbed by the material due to the imaginary part of the dielectric constant which contributes most to heating of the switch. However, in a material used in a HAP laser, it is the modulated beam itself which contributes the most heat. The power absorbed by the switch, P_{abs}, is a product of the incident power, P_o, the absorption coefficient, α, at the appropriate wavelength and the length, l_{prop}, of the material. Given a fixed switching voltage, l_{prop} is fixed by the EO coefficient since $l_{prop} \propto R^{-1}$. Given

$$P_{abs} = P_o \, \alpha \, l_{prop} \tag{2}$$

then

$$P_{abs} \propto \frac{P_o \, \alpha}{R} \tag{3}$$

The quantity to be minimized, when evaluating materials for a HAP EO switch, is α/R. In either case, the heat deposited in the switch generates thermal gradients and is a source of beam distortions due to the dependence of the refractive index on local temperature and stress as well as the dependence of the electro-optic coefficient on the local temperature.

Given the physical properties desired of a HAP EO switch material operating at 1.06 μm, we can derive the corresponding chemical constraints which are imposed.

Physical property desired	Chemical Constraint
Low optical absorption	No hydrogen bonds
No photorefractive effect	No perovskites/refractory melts
Large EO coefficient	?
Good thermal and mechanical properties	Preferred bonding order Covalent-Ionic-Hydrogen-Van der Waals

Optical absorption at 1.06 μm in KDP and other hydrogen bonded materials is dominated by the third overtone of the O-H···O stretch.[3] To eliminate this the material must be deuterated, a procedure which costs ≅ 500$ per liter of D_2O required. Other impurities which have absorptions at 1.06 μm, i.e. Cu and Cr, must be eliminated from the crystal. Materials grown from refractory melts are contra-indicated due to the cost involved in obtaining materials with the minimum impurity levels necessary. In certain structures, notably those with a distorted perovskite structure, these impurities (Fe,Va,Cr) will photoionize. The free electron will polarize the surrounding lattice and stabilize, resulting in a modification of the index of refraction. We will look for materials which are not of the distorted perovskite, $LiNbO_3$ or tungsten bronze type structures. Finally, the material must have good mechanical strength and a high thermal conductivity. It is the type of chemical bonding which contributes most to these properties, with a preferential order being covalent→ionic→H-bonded→van der Waals.

If we examine the list of materials which are available in sizes of 1x10x5 cm^3, we see that the choice of materials for EO switches is extremely limited. Essentially there are only two, listed in Table II, KD^*P and $LiNbO_3$. While these materials have a relatively high absorption coefficient at 1.06 μm compared to SiO_2, a modest switching voltage is achieved for both materials. Beta-Barium Borate (BBO) is a new nonlinear optical material whose use as a HAP EO switch has been previously examined[4]. BBO's attractive features include a low absorption coefficient and extended transparency to the ultraviolet. However, it has a relatively small effective EO coefficient. The switching voltage of a plate of BBO is about 4 times higher than the equivalent size plate of KD^*P or $LiNbO_3$. Finally, it is a flux growth material and is available in cubes about 2 cm on a side.

Quartz represents the ideal EO material in all respects except for one. It is a very durable material, available in large plates with low scattering and absorption losses (at 1.06 μm). It has a high damage threshold and a reasonable heat conductivity. However, it has an extremely small EO coefficient. An SiO_2 plate has a switching voltage 90 times greater than the equivalent KD^*P or $LiNbO_3$ plate.

Table II. Parameters necessary for the evaluation of an EO switch material.

Material	Optical[a] absorption α (m^{-1})	Electro-optic coefficient r (pm/V)	Thermal conductivity K_o(W/m°K)	Damage[a] threshold D_{TH}(J/cm^2)
KD*P	0.4	$r_{xyz} = 24$	2	15
$LiNbO_3$	0.1	$r_{xxx} = 3.1$	0.1	5-7
β–$B\alpha B_2O_4$	0.01	$r_{yyy} = 2.1$	0.8	13.5
Quartz	0.006	$r_{xxx} = 0.23$	5.8	10-12

[a]λ = 1.06 μm, τ = 10 ns

III. WHERE TO LOOK FOR NEW MATERIALS

The EO coefficient can contain contributions from the electronic nonlinearity in conjunction with a contribution due to an ionic modification of the electronic polarizability. The second harmonic generation (SHG) d_{kij} coefficient in general contains contributions only from the electronic nonlinearity. The ratio of r_{ijk} to d_{kij} gives an indication of the size of the relative contribution of the electronic nonlinearity to the EO coefficient.

Upon examination of the origin of the EO coefficients of well characterized materials in Table III, we see that they can be roughly divided into two categories: those which derive their coefficients from the electronic nonlinearity, and those which have a large ionic contribution to the EO coefficient. Materials whose dominant contribution is due primarily to the electronic nonlinearity include quartz, BBO and mNA. If we examine the relative contribution of the SHG coefficient d_{kij} to the EO coefficient r_{ijk}, we see that in these materials this ratio of r/d is approximately 1.

The second group of materials all have large EO coefficients, compared to quartz, but the corresponding d_{kij} coefficient is not necessarily as large. We see that r/d in these materials is very large and the dominant contribution is ionic in nature.

Table III. Relative Contribution of the electronic nonlinearity to the electro-optic coefficient in various materials.

Material	r[a]	d[a]	r/d
Quartz	1	1	1
BaB_2O_4	9.1	5.2	1.8
Urea	7.8	4.6	1.7
mNA	68	42	1.6
SBN	782	35	22
$BaTiO_3$	152	13	11
$LiNbO_3$	3500	19	183
KD*P	105	1.3	83
TGS	234	0.1	2347

[a]normalized to quartz

We can represent the phase space of all EO materials graphically as seen in Figure 1. Moving away from the origin, we find materials with an increasing r_{ijk} coefficient. By following the vertical axis we obtain an EO coefficient by means of the electronic nonlinearity, and by following the horizontal axis we obtain a large EO coefficient by primarily an ionic contribution. A diagonal line drawn at the value of KDP, 10 pm/V, serves as a reference point. Materials from the two groups mentioned above are placed at the appropriate spot on the figure. Materials with large EO coefficients can be found by following either the vertical or horizontal axis away from the origin. However, materials with a large r_{ijk} from the first group obtain it from a large d, which is associated with a lowering of the UV edge and hence increasing optical absorption. The horizontal axis is dominated by H-bonded ferroelectrics and refractory oxides, both groups we have chosen to stay away from.

Examining materials from the middle of Figure I shows us a group of materials with a moderate electronic and moderate ionic contribution to the EO coefficient. Table IV lists a small number of materials from this intermediate category which have an "enhanced electronic" EO coefficient. These materials have a modest r but an even smaller corresponding d, indicating that there is substantial contribution to the overall r_{ijk} coefficient from the ionic contribution.

Table IV. Relative contribution of the electronic nonlinearity to the electro-optic coefficient in several "enhanced electronic" materials.

Material	r^a	d^a	r/d
$Ba(COOH)_2$	8.8	0.3	26
	7.9	0.3	25
$NaNO_2$	8.3	0.2	38
$LiIO_3$	14	0.6	23

[a]normalized to quartz

These materials, which include carbonates, nitrates, formates and iodates, are representative of a group of materials with similar characteristics. They all consist of materials which derive their nonlinearity from small acentric planar (or nearly planar) units which have an enhanced electronic nonlinearity compared to units consisting of tetrahedral units (i.e. phosphates, silicates, etc). Extra enhancement of the r_{ijk} coefficient can occur due to an ionic contribution from a near room temperature phase transition. These materials should all have low optical absorption at 1.06 μm and be free of a photorefractive effect. Finally, these are all low temperature melt growth or solution growth materials, making it relatively easy to obtain large strain free crystals. However, there are no known materials from this group with a large EO coefficient. We hope to find one by looking for a material with a near room temperature phase transition.

The nitrates were the most promising class of materials to examine during our initial study. All nitrates have good chemical stability and are water soluble, facilitating ease of growth. SHG powder measurements indicate that some have a moderate electronic nonlinearity, compared to quartz. Many nitrates also have phase transitions within 150 °C of room temperature. By attempting to make mixed salts of all the binary combinations of single cation nitrate salts, we hope to find new acentric nitrate salts with near room temperature phase transitions. The choice of cations is obvious. Only salts transparent throughout the visible and near IR need be considered. The cations studied included Li, Na, K, Rb, Cs, Mg, Ca, Sr, Ba, Cd, Zn, Pb, Sc, Y, La, Al, Ga, In, Ce and Yb.

The methodology for obtaining new materials for study followed a common procedure. A literature search was conducted on the ternary nitrate phase diagrams; salt A + salt B + H_2O. Many nitrate solubility diagrams exist in the literature due to their uses as fertilizers and oxidizers. Attempts were made (10 cc in volume) to obtain crystals of a mixed salt, if it existed. Powder SHG tests were done on new crystals found to determine acentricity. Those crystals found to be acentric were examined for phase transitions between room temperature and 270 °C. Of those materials deemed interesting, further growth attempts were made to obtain larger crystals (from 100 cc volume solutions) for EO measurements.

The results obtained so far are presented in Table V. The first group of materials listed are those which stand to have a significant ionic contribution due to the proximity of a phase transition "near" room temperature. The second group are materials which may be promising candidates for frequency doubling and whose powder tests indicate substantial electronic nonlinearity, but due to the lack of a phase transition between 0 °C and 270 °C will probably not be good EO switch materials.

IV. MEASUREMENT OF THE EO COEFFICIENTS

The existing experimental techniques for an EO materials survey are either accurate or rapid, but not both. Direct methods include phase modulation or interferometric methods.[4,5]. Both methods require substantial amounts of oriented material (approximately 1cc) in order to accurately determine the EO coefficients. The extension of these techniques to millimeter sized pieces has been done in at least one case on Hexamethylenetetramine (HMT)[6], but this measurement gave a coefficient which was 10 times larger than that measured on centimeter size pieces.[7]

Indirect methods include inferring the EO coefficient from Raman scattering and IR absorption strength data.[8] However, loss of phase information leads to ambiguities in the signs of the contributions. One can also model the EO coefficient using Miller's delta type arguments.[2,9]

Table V. Current results of the search for new EO materials in the phase space of the mixed nitrates.

Material	Point Group	Powder Signal[a]	Trans. Temp. °C
$RbNO_3$	3	3.4	164
$Rb_x(NH_4)_yNO_3$	?	2.4	150-158
$K_x(NH_4)_yNO_3$	?	1.9-0.3	112.6-80.6
$CsNO_3$	3	0.4	154
$Ca(NO_3)_2Urea3H_2O$	mm2	20	deliq.
$BaK_2(NO_3)_4$	?	7.4	>250
$Rb_2Na(NO_3)_3$	mm2	6.0	173[b]
$Cd(NO_3)_24H_2O$	?	4.0	68[b]
$Rb_2Ce(NO_3)_54H_2O$	?	0.2	60[b]
$Cs_2Ce(NO_3)_54H_2O$	?	0.001	64[b]

[a]relative to SiO_2

[b]melts

In 1964, Miller[10] defined a dimensionless coefficient, δ, which correlated a large SHG coefficient with a large index of refraction, n

$$\delta_{SHG} \equiv \frac{\chi^{(2)}}{\left(\chi^{(1)}\right)^3} = \frac{2\,\varepsilon_o\,d}{(n^2-1)^2 n^2}\,. \tag{4}$$

We can define a Millers delta for electro-optic materials:

$$\delta_{EO} \equiv \frac{\chi^{(2)}}{\left(\chi^{(1)}\right)^3} = \frac{2\,\varepsilon_o\,n^4\,r}{4(K-1)(n^2-1)^2}\,. \tag{5}$$

There is some indication that Millers δ is dispersionless throughout the visible frequency spectrum. We assume Millers delta is dispersionless from DC to optical frequencies. Upon equating the two deltas we obtain the relation:

$$r = \frac{-4\,d(K-1)}{\left(n^2-1\right)n^4}\,. \tag{6}$$

The indices of refraction, d_{kij} coefficient, and dielectric constant can be measured or inferred from small (mm size) crystalline samples . But if we examine this relationship using data from the literature[7] in Figure 2, we observe that again we have exchanged ease of measurement for loss of accuracy. The Millers delta model is only good to an order of magnitude.

We propose to use optical rectification as a rapid survey technique for small EO crystals. Given a cube of some acentric material and electric fields of frequency ω and Ω, the following processes can occur. If our input is an optical field of frequency ω, we can observe light of frequency of ω and 2ω at the output. The amplitude of the light of frequency 2ω is directly proportional to the nonlinear dielectric tensor component $\chi^{(2)}(\omega,\omega;2\omega)$. Again, the magnitude of this tensor component is dominated by the electronic nonlinearity. Given an optical carrier of frequency ω and a slowly varying field Ω, we can observe the electro-optic effect, where the optical carrier is modulated by the slowly varying field Ω. The depth of the modulation is directly proportional to the magnitude of $\chi^{(2)}(\omega,0;\omega)$, which contains contributions from the electronic nonlinearity as well as contributions from the ions as they respond to the slowly varying field and affect the electronic nonlinearity. Finally, if the input to our crystal is a modulated light pulse, we can observe the optical rectification (OR) effect, where a voltage pulse whose amplitude follows the modulation of the light pulse appears on the appropriate crystal faces. The amplitude of this voltage pulse is directly related to the magnitude of $\chi^{(2)}(\omega,\omega;0)$, which by symmetry is related to the electro-optic tensor component $\chi^{(2)}(\omega,0;\omega)$,

There are significant advantages of using optical rectification as a measurement technique versus the electro-optic effect. The signal measured is proportional to the input laser power, not the beam intensity. The limit of detection thus depends upon the damage threshold of the material, not the size of the crystal. And unlike the phase modulation technique which measures only the absolute difference between coefficients, optical rectification allows direct measurement of individual coefficients and signs.

We have examined optical rectification in KD*P and quartz in Figure 3. The KD*P sample consisted of a small cylinder (5 mm diameter, 1 cm height) with the z crystallographic axis coincident with the axis of rotation. As we rotate the cylinder about the z axis the 4(bar)2m symmetry of the crystal is observed as the voltage follows the sign of the coefficient r_{xyz}. The cylinder was placed in index matching fluid and the magnitude of the peak voltage was about 10 mV. To test the detection limit, a quartz slab (5 x 5 x 1 cm^3) was measured, where a gold electrode was placed on the (100) faces. If we define ψ as the angle between the x axis and the incident polarization vector, we observe the signal decrease as $\sin^2(\psi)$, as expected. This places the limit of detection approximately around 0.5 pm/V. Finally, preliminary data measured using a combination of OR and phase modulation techniques is presented in Table VI. Several coefficients of DLAP, and $RbNO_3$ have been measured.

Table VI. Preliminary measurements of several new materials.

Material	Coefficient	Value(pm/V)	Coefficient(pm/V))
DLAP	$r_{\beta\beta\beta} - \left[\frac{n_\alpha}{n_\beta}\right]^3 r_{\alpha\alpha\beta}$	1.9 ± 0.2	$r_{\beta\beta\beta} = 2.0 \pm 0.5$
	$r_{\beta\beta\beta} - \left[\frac{n_\gamma}{n_\beta}\right]^3 r_{\gamma\gamma\beta}$	3.5	$r_{\gamma\gamma\beta} = -1.5$ $r_{\alpha\alpha\beta} < 0.5$
$RbNO_3$	$r_{zzz} - \left[\frac{n_o}{n_e}\right]^3 r_{xxz}$	2.1	?

V. CONCLUSION

Currently available EO materials are a limiting factor in the ultimate performance of high average power pulsed lasers. By identifying the physical properties desired with the associated chemical constraints we have identified an area of materials space in which substances with the appropriate physical properties may be found. The phase space of the mixed nitrates is representative of this class and we are exploring it thoroughly. To rapidly survey these materials, new techniques are required to determine the EO coefficients, which do not require large samples. We have proposed and demonstrated that optical rectification has this capability.

VI. ACKNOWLEDGEMENTS

The authors would like to thank Stephan Velsko for his comments and suggestions, both experimental and theoretical. We would also like to thank Charles Barker for his expert advice. This work was performed under the auspices of the U.S. Department of Energy by Lawrence Livermore National Laboratory under Contract No. W-7405-ENG-48.

VII. REFERENCES

1. I. P. Kaminow, An Introduction to Electro-optic Devices, (Academic Press Inc., 1974).
2. D. Eimerl, Ferroelectrics, 72, 95 (1987).
3. C. A Ebbers, S. Velsko, "Absorption at 1.06 μm in highly deuterated KDP", LLNL memo. LRD 88-204/6123K.
4. C. A. Ebbers, "β–Barium Borate as an electro-optic material for HAP lasers", SPIE Vol. 968, (Aug. 15-16, 1988).
5. M. Sigelle, R. Hierk, J. Appl. Phys., 52, 4199, (1981).
6. R. W. McQuaid, Appl. Opt., 2, 320, (1963).
7. W. R. Cook Jr. and H. Jaffe, Electro-optic Coefficients, Landolt-Bornstein, New Series, K.-H. Hellwege, ed., Vol. 11, (Springer-Verlag, Berlin,1979) p. 556.
8. I. P. Kaminow, W.D. Johnston Jr., Phys. Rev. ,160, 519 (1967).
9. M. A. Henesian, "An empirical formula for predicting Electro-optic and Second Harmonic Coefficients", LLNL memo. UVM 84-09/4012T.
10. R. C. Miller, Appl. Phys. Lett., 5, 17, (1964).

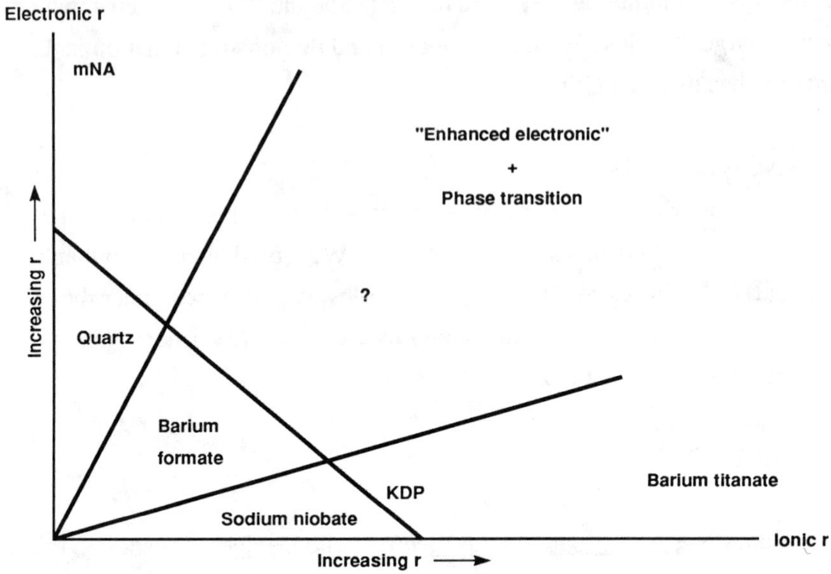

Figure 1. Electro-optic materials grouped according to the dominance of the electronic or ionic contributions.

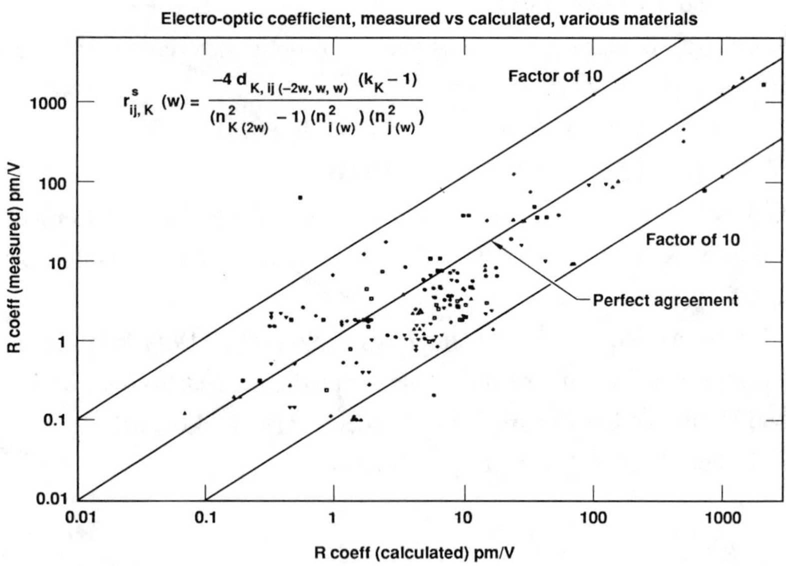

Figure 2. Miller's δ scaling for EO materials is good to an order of magnitude.

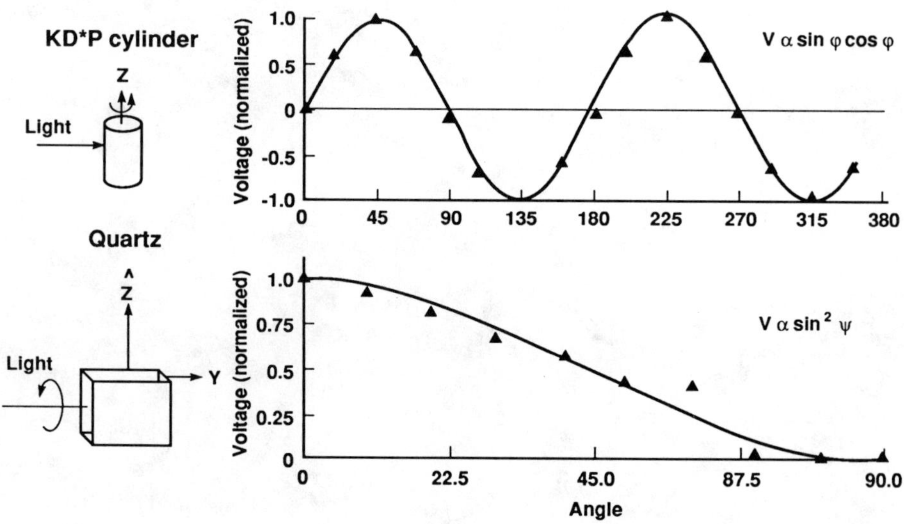

Figure 3. Observations of optical rectification in a cylinder of KD*P and quartz.

SOME PROPERTIES OF $Li_{1-X}Nb_{1-X}Eu_{2X}O_3$ SOLID SOLUTIONS

M. E. Villafuerte-Castrejón

Instituto de Investigaciones en Materiales, Universidad Nacional Autónoma de México, P.O. Box 70-360, 04510 México, D. F.

A. R. West

Department of Chemistry, University of Aberdeen Meston Walk, Aberdeen, AB9 2UE Scotland

and

J. O. Rubio

Departamento de Física, Universidad Autónoma Metropolitana Iztapalapa, P. O. Box 55-534, 09340, México, D.F.

ABSTRACT

In the present paper the effects of europium substitution on crystalline solubility, structural changes and optical properties of $LiNbO_3$ is reported. The solid solutions were characterized by X-ray powder diffraction and density measurements. The analysis of the optical data revealed that two equally abundant europium luminescent species exist in the stoichiometric specimens; these correspond to the Eu^{3+} ions occupying the Li^{+} and Nb^{5+} sites, in agreement with the replacement mechanism which was inferred from density measurements. Other Eu^{3+} luminescent species were found to exist in nonstoichiometric samples in addition to those which are present in the stoichiometric materials.

INTRODUCTION

Since the influence of lattice defects on the physical properties of $LiNbO_3$ is a subject of considerable interest, different kinds of doubly valent (Mn, Mg, Cu, Co, Zn, Fe) and trivalent (Fe, Gd, Eu, Er, Nd, Dy, Yb) impurity ions have been introduced into the crystal structure of this compound. It is well known, that the Curie temperature of $LiNbO_3$ is very sensitive to the substitution cation localized in either the octahedral or tetrahedral sites of the lattice. Therefore, a reasonable knowledge of the sites occupied by the added impurities in the lattice of this compound is quite fundamental for obtaining a good understanding of the physical properties of doped-lithium niobate. However, this knowledge is in most cases, rather difficult to obtain partly due to the pseudo-ilmenite structure[1-2] of $LiNbO_3$ in which three possible sixfold coordinated axial sites are available for impurities; a Li site, a Nb site and a structural vacancy, all of them being characterized by the same point-symmetry group (C_3). Due to this situation, an unequivocal determination of the sites occupied by divalent and trivalent impurities in $LiNbO_3$ is difficult and has been the subject of controversy over several years.

A previous study[3] has been carried out on the optical emission and excitation spectra of the Eu^{3+} ion in single crystals of $LiNbO_3$. It was established from the analyses of the optical spectroscopic data, that the Eu^{3+} ions enter both the Li^+ and Nb^{5+} sites. The finding of only two different Eu^{3+} luminescent species in this lattice is quite surprising especially since the available information was obtained from single crystals which were grown with the congruent melting composition. It is well known that such a composition implies a large number of defect centres with an uncertain nature. Moreover, impurity solubility was not determined previous to the interpretation of the experimentally determined data. Therefore, the impurity ions are expected to occupy several kinds of different sites due to the several ways in which other defects introduced by the lithium deficiency and possible excess of impurity concentration may be associated with it.

In view of the increased interest in Eu^{3+}-activated oxides for their possible use as red phosphors in color television tubes and fluorescent lamps, it may be rather interesting: a) to analyze the optical properties of Eu^{3+} in stoichiometric specimens of $LiNbO_3$, b) to determine the solubility ranges of the trivalent europium ion in this compound and c) to get a better insight into the intriguing finding reported in Ref 3. Therefore, the purpose of the present paper is to present a detailed experimental study of the effects of Eu^{3+} substitution on crystalline solubility, structural changes and optical properties of $LiNbO_3$.

EXPERIMENTAL

All measurements described in this paper were performed on powder samples which were synthesized by solid state reaction. Starting materials were Li_2CO_3 (Baker reagent), Nb_2O_5 (99.9% Aldrich) and Eu_2O_3 (99.9% Norton Thiokol, Inc.). Mixtures totalling 3 to 5 g were prepared by weighing, mixing into a paste with acetone in an agata mortar, and finally fired in Pt crucibles. Initial firing was at 973 K for a few hours to expell CO_2 and finally at 1423 K for 3 to 7 days. All reactions were conducted in air under atmospheric pressure. Weighing control before and after reaction allowed us to check the complete loss of CO_2, the composition stability of the sample and that lithium loss by evaporation did not occur to an appreciable extent. Only preparations with weight loss lower than 0.2 mole % were considered for the chemical and spectroscopic analyses. The completeness of reaction and phase indentification was checked through X-ray diffraction analyses of the sintered powders. Densities were measured with specific gravity bottles using CCl_4 as displacement liquid.

Optical spectroscopic measurements were performed on white samples having the shape of disks (~5 mm in diameter and ~0.5 mm in thickness) using a conventional set-up consisting of an Oriel 150-W Xenon lamp as light source and two Pacific 0.5m Czerny-Turner monochromators.

RESULTS

a) CRYSTAL CHEMICAL DATA

The data shown in the ternary diagram portrayed in Figure 1, shows that lithium niobate forms with Eu_2O_3 a very limited range of solid solutions in which the replacement mechanism seems to be given by:

$$Li^{+} + Nb^{5+} \rightleftarrows 2Eu^{3+}$$

According to this mechanism, the solid solutions are described by:

$$Li_{1-X}Nb_{1-X}Eu_{2X}O_3 \qquad 0 < X \leq 0.01$$

Beyond the limit of solubility of X = 0.01, the ceramic samples showed extra reflections in the X-ray diffraction patterns, in addition to the $LiNbO_3$ -type ones, implying the coexistence of a different phase with the formula $EuNbO_4$.

The variation of density (ρ) of the europium-doped $LiNbO_3$ specimens was measured within the range of compositions of the solid solutions. The results obtained are given in Figure 2. Density values are accurate within ± 0.05% In the same figure the calculated values of ρ using the solid solution mechanism Li^{+}+ Nb^{5} $2Eu^{3+}$, in which the total number of cations is constrained to remain constant are also included for the sake of comparison. The reasonable agreement between experimental and calculated magnitudes of ρ confirms the mechanism of solid solution formation in which both Li^{+} and Nb^{5+} ions are substituted by Eu^{3+} ions.

Figure 2 also shows the changes in the hexagonal lattice constants and c/a ratio of the pseudo-ilmenite solid solutions as a function of europium concentration.

In addition to the solid solutions which lie on the stoichiometric join $LiNbO_3$-Eu_2O_3, an area of both Li-rich and Li-deficient solid solutions forms, as indicated in Fig. 1b. A characterization of this area has not been made since this determination was not the main objective of the present investigation.

b) OPTICAL DATA

Within the range of compositions of $Li_{1-X}Nb_{1-X}Eu_{2X}O_3$ solid solutions, the luminescence of the Eu^{3+} ions was found to be extremely intense and increased with concentration. Figure 3 portrays, as an example, the 300K emission pattern obtained from a $Li_{0.99}Nb_{0.99}Eu_{0.02}O_3$ specimen under different excitation wavelengths. The strongest lines are in the red region of the electromagnetic spectrum originating from the $^5D_0$ level.

The excitation spectrum corresponding to the prominent $^5D_0 \rightarrow {}^7F_2$ emission transition, consists of a number of sharp lines which are due to transitions within the Eu^{3+} 4f shell. A broad band located in the U.V. region of the electromagnetic spectrum was also present in the excitation spectrum. It was associated with an Eu^{3+}-O^{2-} charge transfer band[4-6].

Similar results to those described above were obtained in the $Li_{1-X}Nb_{1-X}Eu_{2X}O_3$ solid solutions containing a smaller concentration of Eu^{3+} ions.

In order to obtain data which might give an indication of the site occupied by the europium ion in the lattice of $LiNbO_3$, the second derivative presentation of the emission spectrum shown in Figure 3 was recorded under the very high resolution conditions of our experimental set-up. Figure 4 displays the results obtained for the group of lines corresponding to the $^5D_0 \rightarrow {}^7F_0$ and $^5D_0 \rightarrow {}^7F_1$ emission transitions.

In an attempt to learn about the influence of Li/Nb ratio on the luminescence properties of europium-doped $LiNbO_3$, several lithium deficient (nonstoichiometric) samples were also prepared and optically analyzed. Figure 5 shows, as an example, the 80K emission pattern obtained from a $Li_{0.97}Nb_{1.01}Eu_{0.02}O_3$ specimen under different excitation wavelengths. These spectra are quite different to those portrayed in Figure 3. Their intensities increased with the lithium deficiency suggesting, therefore, that several Eu^{3+} fluorescent centres, different to those existing in the stoichiometric samples, were created by the lithium deficiency.

The peak position and shape of the broad charge transfer band present in the excitation spectrum of the Eu^{3+} luminescence in the lithium deficient samples were found to be quite sensitive to the emission wavelength which was selected to record the excitation spectrum as can be appreciated from the data portrayed in Figure 6; this finding being a clear indication that different Eu^{3+} luminescent species whose emission spectra were strongly superimposed existed in the lithium deficient specimens.

DISCUSSION

The emission spectra shown in Figure 3 for the $Li_{0.99}Nb_{0.99}Eu_{0.02}O_3$ stoichiometric sample mainly consist of the electric dipole transition $^5D_0 \rightarrow {}^7F_2$ which is parity forbidden. This suggests that the crystallographic sites occupied by the Eu^{3+} ions in the $Li_{1-X}Nb_{1-X}Eu_{2X}O_3$ ($x \leq 0.01$) solid solutions lack inversion symmetry. This conclusion can be supported by the observation of the $^5D_0 \rightarrow {}^7F_0$ transition in the emission spectra of the stoichiometric powders. This transition is forbidden but it might be observed if the Eu^{3+} ion occupies the Li^+ and/or Nb^{5+} sites in the lattice of $LiNbO_3$ since both of them are described by the same C_3 point symmetry group. A better insight into this situation can be gathered from the spectroscopic data described above. In particular, since the $^5D_0 \rightarrow ^7F_0$ transition can not be split by any crystal field, the existence of two emission bands in the high resolution second derivative presentation of this emission may be taken as strong evidence that two equally abundant Eu^{3+} fluorescent species existed in the $Li_{1-X}Nb_{1-X}Eu_{2X}O_3$ ($x \leq 0.01$) solid solutions and they were responsible for the series of strongly superimposed luminescence transitions. This conclusion can be also supported by the following facts: (a) In an hexagonal site symmetry two peaks associated with the $^5D_o \rightarrow ^7F_1$ transition are expected to be resolved in the emission spectrum. Our data in Figure 4 clearly show that this transition displays four lines with nearly the same intensity indicating therefore that two equally abundant species of Eu^{3+} ions existed in our stoichiometric specimens and (b) The second derivative presentation of the charge transfer transition between the Eu^{3+} and O_2^- ions revealed that this band was really composed of two strongly superimposed excitation bands; each one of these two bands being associated with a different Eu^{3+} luminescent centre in the lattice of $LiNbO_3$.

Three similar sites are available for cationic impurities in the lattice of $LiNbO_3$, a Li^+ site, a Nb^{5+} site and a structural vacancy. On the basis of charge differences the Eu^{3+} ion is as likely to enter the Li^+ sites as the Nb^{5+} sites but less likely to enter the vacancy sites. From our optical data, it might be concluded that, within the range of compositions of the solid solutions, the Eu^{3+} ions occupy both the Li^{+} and Nb^{5+} sites in the lattice of $LiNbO_3$.

The data shown in Figure 2 show that, the unit cell volume of $LiNbO_3$ increases with the introduction of Eu^{3+} ions. This result is expected since the radius of the Eu^{3+} ion is bigger than those of Li^{+} and Nb^{5+}. This increase in the unit cell volume does not produce significant changes in neither the emission nor in the excitation spectra of the europium-doped stoichiometric specimens. This fact may be explained considering that the 4f electrons in a rare earth ion are shielded from the crystal field. Therefore, one may expect that the 4f→4f transition energies are not very sensitive to crystalline changes.

Finally, the most significant results obtained in the present investigation can be summarized as follows:

a) Lithium niobate forms with Eu_2O_3 a very limited range of solid solutions in which the replacement mechanism seems to be given by $Li^{+}+Nb^{5} \rightleftarrows 2Eu^{3+}$.

b) Within the range of compositions of the solid solutions, the unit cell volume of the pseudo-ilmenite phase of $LiNbO_3$ increases with increasing Eu_2O_3 concentration.

c) Two equally abundant europium luminescent species were found to exist in the stoichiometric specimens of lithium niobate. These correspond to the Eu^{3+} ions occupying the Li^{+} and Nb^{5+} sites in agreement with the replacement mechanism which is inferred from density measurements.

d) Other Eu^{3+} luminescent species were found to exist in the nonstoichiometric lithium-deficient samples, in addition to those which were present in the stoichiometric materials. The extra luminescent species appear to be closely associated with the defects which are created by the lithium deficiency. Unfortunately, the nature of these additional Eu^{3+} luminescent centres can not be inferred from our data.

The previous finding[3] of only two different Eu^{3+} luminescent species in the single crystals of $LiNbO_3$ which had been grown with the congruent melting composition is quite in contrast with the spectroscopic evidence presented in this paper. Moreover, the europium concentration mainly employed in this previous work to obtain the optical data is, according to our determination, well outside the limits of solubility. Under these circumstances, it is expected that the Eu^{3+} ion may find itself in several kinds of different sites due to several ways in which it may be associated with other defects introduced by both the lithium deficiency and excess europium concentration. This conclusion is supported by our spectroscopic optical data.

We may conclude that the data presented in this paper have provided a considerable amount of insight into the effects of Eu^{3+} substitution on crystalline solubility, structural changes and optical properties of lithium niobate. They also emphasize the importance of performing chemical and spectroscopic measurements in parallel for a reasonable characterization of the sites occupied by the added impurities and an understanding of defect-structure specific property relations.

ACKNOWLEDGEMENTS

This work was partially supported by Consejo Nacional de Ciencia y Tecnología (CONACyT) and British Council through the Aberdeen-Mexico collaboration programme. The authors acknowledge I. Camarillo and L. Baños for the technical assistance.

REFERENCES

1.- Abrahams, S.C., Hamilton W.C. and Ready J.M., 1966a J. Phys. Chem. Solid 27. 1013; 27 1019; Abrahams, S. C. Reddy, J. M. and Berstein, J. L. 1966b, J. Phys. Chem. Solis 27, 997

2.- Megaw H., 1968 Acta Cryst. A24, 583

3.- Arizmendi L and Cabrera J.M. 1985, Phys. Rev. B31, 7138

4.- Blasse G and Brill 1966 J. Chem. Phys. 45, 3327

5.- Blasse G in Luminescence of Inorganic Solids 1978 pag 457 Ed. B DiBartolo Plenum Press, New York.

6.- Hoefdraad H. E., 1975 J. Sol. St. Chem. 15, 175

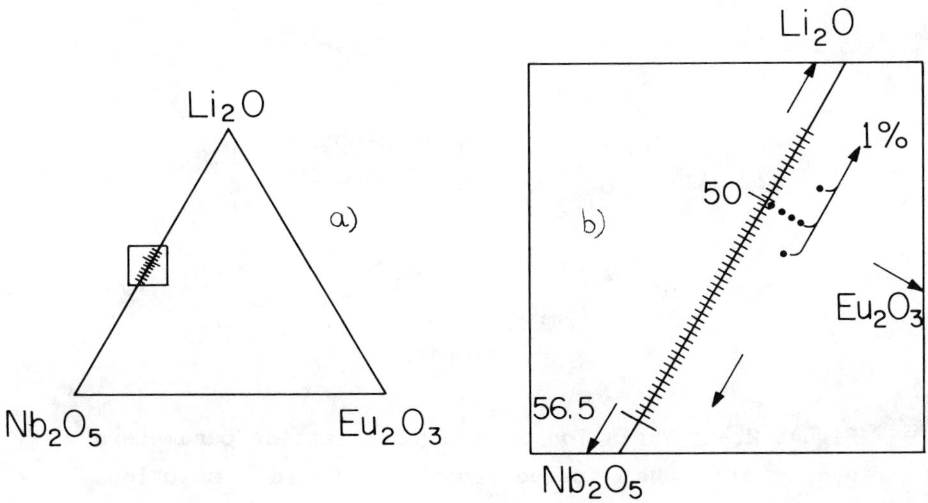

Figure 1.- Solid solutions in the system Li_2O-Nb_2O_5-Eu_2O_3 which lie on the stochiometric join $LiNbO_3$-Eu_2O_3. An area of both Li-rich and Li-deficient solid solutions forms, as indicated in Fib. 1b.

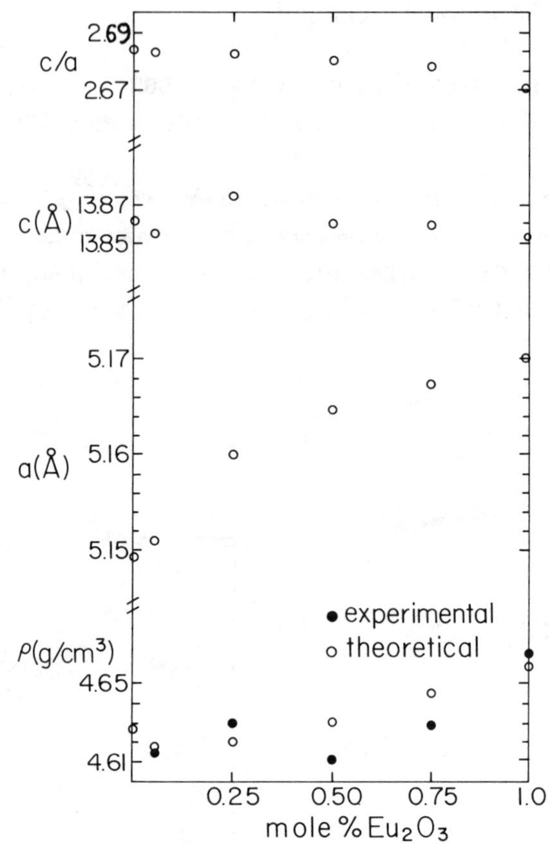

Figure 2.- Variation of hexagonal lattice parameters a, c, of the pseudoilmenite solid solutions and c/a ratio as a function of Eu_2O_3 concentration. Experimental and calculated densities (ρ) of solid solutions are included. The bars in this figure represent experimental uncertain ranges.

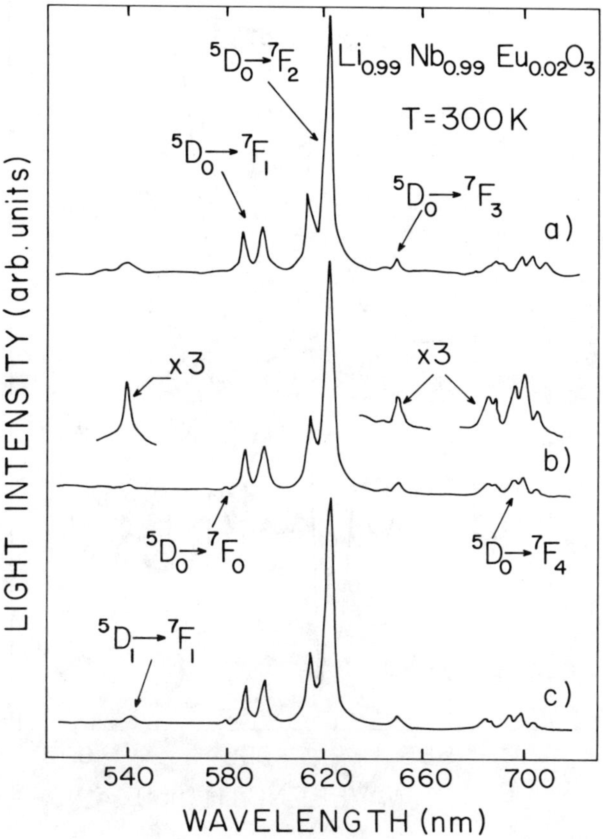

Figure 3.- Room temperature emission spectrum of $Li_{0.99}\ Nb_{0.99}\ Eu_{0.02}O_3$ as a function of different excitation wavelengths: a) 270 nm, b) 300 nm and c) 399 nm.

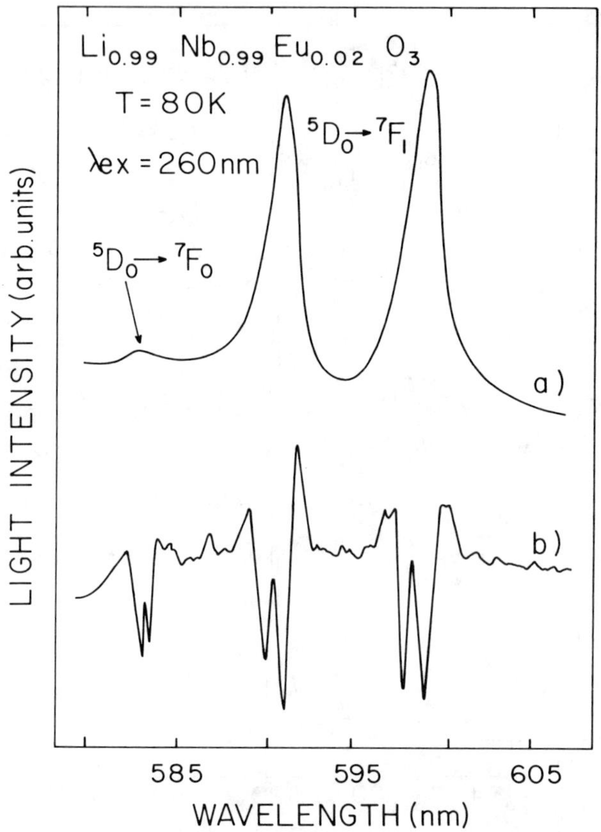

Figure 4.- a) Liquid nitrogen emission spectrum showing the $^5D_0 \rightarrow {}^7F_0$ and $^5D_0 \rightarrow {}^7F_1$ transitions of Eu^{3+} ions in stoichiometric $Li_{0.99}Nb_{0.99}Eu_{0.02}O_3$ specimens; b) second derivative presentation.

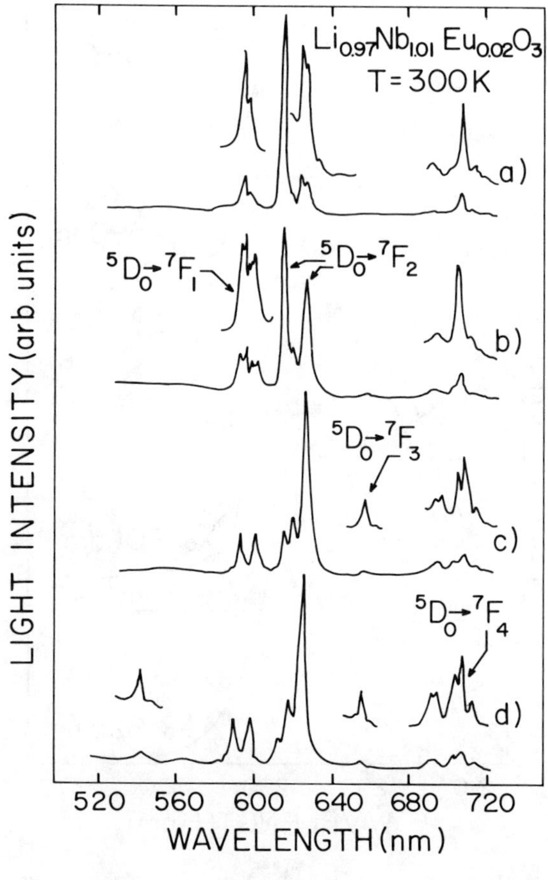

Figure 5.- Room temperature emission spectrum of a lithium deficient $Li_{0.97}Nb_{1.01}Eu_{0.02}O_3$ specimen as a function of different excitation wavelengths: a) 260 nm, b) 270 nm, c) 300 nm and d) 398 nm

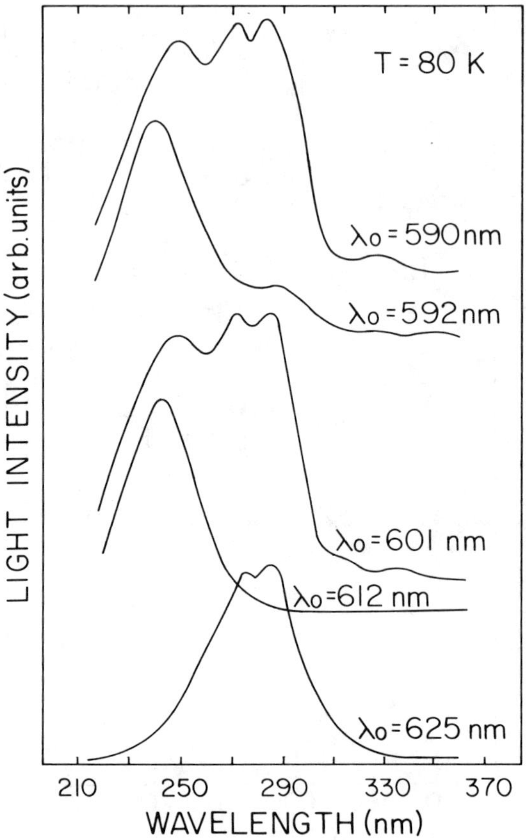

Figure 6.- Liquid nitrogen charge transfer excitation spectrum as a function of different emission wavelengths in lithium deficient $Li_{0.97}Nb_{1.01}Eu_{0.02}O_3$ specimens.

Section II. Second and Third Order Nonlinear Optical Effects

Nonlinear Effects In Optical Fibers

Mansoor A. Saifi
Bellcore
331 Newman Springs Road
Red Bank NJ. 07701

Nonlinear optics may be defined as the effects associated with the nonlinear response of atoms and molecules to optical radiation fields. In general for an intense electromagnetic field the response of any dielectric medium becomes nonlinear, due to anharmonic motion of bound electrons. The general expression of electric field induced polarization **P**, which includes both the linear and the nonlinear response, is given by:

$$\mathbf{P} = \epsilon_0 \left[\chi^{(1)}E + \chi^{(2)}E\,E + \chi^{(3)}E\,E\,E + \ldots \right]$$

where ϵ_0 is the free space permittivity and $\chi^{(j)}$ jth order susceptibility. In addition to their dependence on crystallographic direction the optical susceptibilities vary with the polarization direction of the light. For this reason $\chi^{(j)}$ is a tensor of rank (j+1). Most of the nonlinear effects, in optical fibers, are related to second and third order susceptibility. Furthermore, stimulated Raman and stimulated Brillouin scattering are two other nonlinear effects of considerable interest for optical communication systems.

Nonlinear effects in optical fibers can provide useful device functions, such as optical-frequency conversion, optical amplification, and optical signal processing. The same effects also impose limits on the transmission system due to undesirable signal loss, signal distortion and cross-talk in adjacent channels. The purpose of this paper is to review nonlinear effects in single mode optical fibers and their relation to fiber parameters. In particular the review will focus on a few second and third order nonlinear effects and stimulated scattering processes. Nonlinear fiber optics has seen intense activity during the past fifteen years and it has been a subject of several previous reviews (1-3). Recently a book dealing exclusively with this subject (4) has appeared.

Optical Fiber Key Features For Nonlinear Effects

The nonlinear effects we will review are in high silica single mode fibers doped with one dopant or a combination of dopants, particularly oxides of germanium, phosphorous, titanium, and boron. Compared to crystalline dielectric materials such as

KTP, $LiNbO_3$, ADP, the nonlinear susceptibilities in high silica glasses are lower by at least one or two orders of magnitude. In spite of this, many of the nonlinear effects in single mode fibers have been observed at moderate to high power levels. This success can be attributed to three key features of single mode optical fibers. First, the small core diameter, few microns, implies high optical power density. Second, the guiding structure results in long interaction length. Third, the material and the waveguide dispersion can be combined to satisfy certain phase matching conditions. These features are illustrated in Fig.1. In bulk materials high field intensities are achieved by focusing the light. However, diffraction limits the effective interaction length to at most a few mm. For a gaussian beam the effective interaction length is given by $L_{eff} = \pi a^2/\lambda$, where a is the focused spot radius and λ the wavelength of the light. The product, intensity I x the effective interaction length, is a good measure or figure of merit for nonlinear optics. For a gaussian beam in bulk optics,

$$(IL_{eff})_{bulk} = (P/\pi a^2)(\pi a^2/\lambda) = P/\lambda ,$$

is constant. Therefore, tighter focusing does not help to enhance the nonlinear process. On the other hand, for a fiber with an absorption coefficient of $\alpha(cm^{-1})$, the effective interaction length is $1/\alpha$ cm. The product IL_{eff} in fiber is,

$$(IL_{eff})_{fiber} = P/\pi a^2\alpha$$

Comparing with bulk optics, the efficiency of the nonlinear process in fiber is enhanced by a factor of,

$$= (IL_{eff})_{fiber}/(IL_{eff})_{bulk} = \lambda/\pi a^2\alpha$$

For a typical single mode fiber, with a mode field radius of 5 μm, and a loss of 2 dB/km ($\alpha = 5\times10^{-6}$ cm^{-1}) at 1 μm wavelength, the enhancement factor exceeds 10^{10} ! Here then lies the key for the success of nonlinear effects in optical fibers. However, as will become evident, considerable improvements in the nonlinear optical coefficients of the glass are required for practical devices using semiconductor lasers as light source.

Another key feature of nonlinear optics in fibers is the ability to tailor the overall chromatic dispersion by suitable design of the fiber index profile and core dimensions. In a fiber the dispersion term contains contribution from both the material and the waveguide dispersion. The waveguide dispersion depends on the core/clad index difference, core diameter and index profile. For example in a standard single mode fiber, the group velocity dispersion (GVD) is positive for wavelengths shorter than 1.32 μm, zero near 1.3 μm, and negative for longer wavelengths. The zero dispersion wavelength and the corresponding overall dispersion curve can be shifted to longer wavelengths by reducing the fiber core diameter and if necessary manipulating the core clad index difference, see Fig.2. This facility to tailor the dispersion is utilized to observe effects such as pulse compression, soliton transmission, and four photon mixing.

Second Harmonic Generation (SHG)

Efficient second harmonic generation (SHG) has been observed only recently in optical fibers. Sum-frequency light generation in glass fibers is a well-known phenomenon (5-7), however, efficient SHG discovered by Osterberg and Margulis (8) is an unusual effect. This is because second order processes are normally dipole forbidden in materials with a center of inversion. Amorphous materials do not posses an ordered structure, in particular long-range order. Glass fibers, therefore, would normally not exhibit a strong second-order optical susceptibility. The observation of unusually efficient SHG, over 5% conversion efficiency (9), is due to certain unique features. At the outset we point out that, there are several higher order nonlinearities such as quadrupole electric and magnetic-dipole moments and nonlinearities at the core/clad interface (10), which would allow SHG. However calculations show that these could result in SHG conversion of at most 10^{-5}, whereas observed conversion is several orders of magnitude higher.

The Osterberg and Margulis (8,9) discovery centers on irradiating standard telecommunication-grade single-mode fiber (<50 cm to few meters long) for several hours with high intensity (40-kW peak power) pulses from a mode-locked and Q-switched Nd:YAG laser, operating at 1064 nm. As shown in Fig.3, during this long exposure, the generated second harmonic builds up by nearly six orders (from nanowatts to milliwatts) of magnitude. Once prepared in this manner, a nearly permanent change is produced in the fiber and frequency doubling occurs even at comparatively low input power levels, with conversion efficiency approaching as high as 5%. Figure 3 clearly shows that SHG build up accelerates with second harmonic power in the fiber. The process may initiate from existing defects as was observed in the earlier work (5,6).

Several models have been proposed to explain this phenomenon. A common theme amongst the proposed models is formation of a phase matching grating to provide long interaction length between the fundamental and the second harmonic polarization wave. In the Stolen and Tom (11) model, the pump light at 1064 nm and the 532 nm harmonic light mix through the third-order optical susceptibility, generating a spatially alternating dc polarization field. The production of dc field by third-order process is analogous to second-order optical rectification (12). The third-order dc field is given by,

$$E_0 = \mathrm{Re}\,[\,\chi^{(3)} x\; E_1^* \, E_1^* \, E_2 \exp i(k_2 - 2k_1)]$$

where E_1 and E_2 are fundamental and second harmonic field intensities, and k_1, k_2 the respective wave vectors. The wave vector mismatch $\Delta k=(k_2-2k_1)$ produces spatially alternating dc fields of period $2\pi/\Delta k$ (30 to 40 μm for 1060 nm fundamental). The spatially alternating dc field , with accompanying distribution of charges, shown in Fig.4, breaks the inversion symmetry, and has proper periodicity for phase matched SHG. Furthermore the induced second-order nonlinearity is proportional to the dc field.

$$\chi^{(2)} = \eta E_0$$

Moreover, Stolen and Tom (11) showed that the fiber preparation time can be reduced from several hours to several minutes, by initiating the process with an externally generated second harmonic light. For this they used an experimental set up similar to Fig.5, and illuminated the fiber simultaneously with both the fundamental and its second harmonic generated by a KTP crystal.

There is yet another model by Farries et.al (13), which also proposes a self-generated alternating $\chi^{(2)}$ grating. In this model, color centers are generated by a second-order process in the regions where the 532 nm radiation, initially from quadrupole generation, is most intense.

A criticism (14) of the Stolen and Tom model is that the generated dc field is of the order of 1 v/cm, too weak to induce permanent $\chi^{(2)}$. However, the fact that a grating is generated has been experimentally verified. In one experiment this was done by measuring the wavelength dependence of second harmonic conversion efficiency (13), and in another experiment by detuning (15) the grating by stretching the fiber.

The above models do not elucidate the physical entity participating in the dipole formation. Initially it was suggested that phosphorous doping is essential (16) in observing efficient SHG. This has been shown not to be the case (17) and in fact efficient SHG has been achieved in silica fibers doped with germanium only. In addition, recently Tsai and co-workers (18) observed a correlation between SHG efficiency and Ge E′ defect centers in fibers prepared for efficient SHG. These and other defects may be the source of the observed second order nonlinearity.

To summarize then, efficient SHG in glass fibers is an unusual phenomenon. That a grating is formed is well established, though the microscopic details regarding its formation mechanism and the nature of physical entity responsible for it, are yet to be worked out. The possibility that $\chi^{(2)}$ is related to electronic defects in fibers has sparked considerable interest in the defect structure of glass.

Third Order Nonlinear Effects

Third order nonlinearity, $\chi^{(3)}$ is also described as the intensity dependent refractive index change. This can be seen from the following.

$$\mathbf{D} = \epsilon E = \epsilon_0 E + \mathbf{P}$$

$$n^2 = \epsilon/\epsilon_0 = [(1+\chi^{(1)})E + \chi^{(3)}I]$$

where ϵ/ϵ_0 is the dielectric constant at optical frequencies and is equal to the square of the refractive index, and E^2 is replaced by the power density I. For $\chi^{(3)} << (1+\chi^{(1)})$, the above can be approximated by (4),

$$n = n_0 + n_2 I$$

where n_0 is the linear refractive index, n_2 the intensity dependent refractive index and $n_2 I$ the light intensity induced refractive index change. The intensity dependent change in the refractive index gives rise to many interesting nonlinear effects in optical fibers. In the following we discuss three such effects, optical switching, the optical Kerr effect, and self phase modulation.

Optical Switching

The first effect due to $\chi^{(3)}$ which has attracted considerable recent attention, is optical switching in a nonlinear directional coupler reported by Frieberg et.al (19). The linear directional coupler of Fig.6 is constructed with a twin core single mode fiber. The propagation constants of the two cores are exactly matched, and the separation between the two cores is such that their evanescent fields overlap. The coupling due to overlap in the evanescent fields, produces periodic transfer of energy between the two cores. This is similar to periodic transfer of energy between two coupled pendulums. The power at any length L in core 2 for light coupled at the input of core 1 is given by the following expression.

$$I_2(L) = I_1(0)\ \mathrm{Sin}^2(CL/L_c)$$

In the above L_c is defined as one coupling length, and C is the coupling coefficient. The coupling coefficient C depends on separation between the two cores and their propagation constants. The coupling length L_c is the minimum coupler length for which at low intensities all the light coupled into core 1 appears at the output of core 2. A coupler of length L_c is called a linear directional coupler. In a directional coupler, the coupling between the two cores is reduced if their propagation constants are not matched. This again is similar to coupling between two pendulums with differing resonant frequencies, which results in an incomplete transfer of energy. In a nonlinear directional coupler, the switching of power at the output from core 2 to core 1 is achieved by changing the propagation constant of core 1. This is brought out by the intensity dependent change in its refractive index (20). The critical power P_c at which complete switching is achieved is given by, $P_c = \lambda A/n_2 L_c$, where A is the core area. Rearranging this gives,

$$P_c n_2/A = \lambda/L_c\ .$$

The left hand side of the above is the intensity dependent change in the refractive index needed to cause switching. Note that at a given wavelength the switching power is lower for a coupler of longer length L_c. For example the index change required for a coupler of 1 cm length at 1 μm wavelength is about 10^{-4}. This coupler, fabricated from a silica core ($n_2 = 3.2\times10^{-16}\ cm^2/W$) single mode fiber of 5 μm core diameter will require a switching power of about 60 kW.

Such high powers are not available from semiconductor lasers. The switching experiments reported by Freiberg (19) et.al were therefore conducted with pulses from a mode-locked Q-switched Nd:YAG laser operating at 1064 nm. Switching power can be reduced by using couplers of sufficiently long length. For example, at a wavelength of 1 μm, the switching power will be 600 watts for a 1m long coupler, requiring a light induced refractive index change of 10^{-6}. This means the refractive indices of the two cores should be controlled to better than the light induced change of 10^{-6}. Thus the required fabricating tolerances are so tight as to make this option impractical. Several other schemes have been proposed to lower the switching power (21,22), but so far none have proved practical. Another approach is to enhance the material's nonlinear optical coefficient. However, materials with higher nonlinear optical coefficient seem to show higher linear and/or nonlinear (two photon) absorption (23,24). An extensive discussion on the guide geometry and material considerations for nonlinear optics can be found in a review article by Stegaman and Stolen (25).

Another problem with all nonlinear switching devices is that, switching occurs, only for that portion of the pulse where intensity is above the critical value. This causes pulse break up as illustrated in Fig.7, which has been experimentally confirmed (26). The switching can be improved by using nearly square shaped pulses as reported by Weiner et.al (27). Another option is to use solitons, which due to their unique features can be switched completely (28). The special nature of optical solitons is described in a later section.

Optical Kerr Effect

The change in the refractive index with the square of the applied electric field and higher order even terms is generally referred to as the quadratic electro-optic effect or the Kerr effect. In analogy, light intensity induced birefringence in optical fibers, is termed, optical Kerr effect. This effect can be used to change the transmission of a weak probe through the birefringence induced by a strong pump. The operating principle of optical Kerr effect can be understood by referring to Fig.8. The pump and the probe beam are linearly polarized at the input end. The pump is polarized along one of the principal axes of a birefringent (30) fiber, and the probe at 45° to this axis. In this configuration, the pump polarization remains fixed, and is blocked by the crossed polarizer at the output. On the other hand the probe polarization rotates, and at the output its component parallel to the crossed polarizer gets detected. The birefringence experienced by the probe depends on the pump intensity and the corresponding refractive index change along the pump polarization direction. In an optical Kerr shutter, in the absence of pump, initial probe polarization is such that it is blocked by the crossed polarizer at the output end. Injecting the pump beam of proper intensity rotates probe polarization by 90°, which allows it to pass through the crossed polarizer. The optical Kerr shutter speed could be as fast as a few picoseconds. It is apparent that by modulating the pump intensity, this effect can be used to modulate the probe. In fact the long interaction length means only modest pump powers suffice for 100% modulation of the probe. For example, in a 500 meter long fiber , nearly 100% modulation of 1150 nm probe was achieved with 390 mW of 1060 nm pump power (29).

Self-Phase Modulation (SPM)

Self-phase modulation (SPM) can be described as spectral broadening of an optical pulse due to the intensity dependent change of a media's refractive index. A general description of SPM in optical fibers requires a numerical solution of the pulse propagation equations with nonlinear refractive index (31,32). Following gives a simple description to illustrate the basic phenomenon of SPM. To this end, the phase of a plane propagating wave of frequency ω_0, and wavelength λ is given by,

$$\phi = \omega_0 t - (2\pi/\lambda) n_0 z$$

where n_0 is the medium's linear refractive index. In the linear regime and for constant z, the time derivative of the phase gives the frequency of the propagating wave. The inclusion of the intensity dependent refractive index gives,

$$\phi = \omega_0 t - (2\pi/\lambda)(n_0 + n_2 I)z$$

Once again the phase excluding the constant term due to n_0 is ,

$$\Delta\phi = \omega_0 t - (2\pi/\lambda) n_2 I z$$

In this case the instantaneous frequency, for length z, is given by,

$$d(\Delta\phi)/dt = \omega_0 - (2\pi/\lambda) n_2 z (dI/dt)$$

Thus, the inclusion of the intensity dependent refractive index change results in an additional frequency term proportional to $n_2 z(dI/dt)$. This means that for the leading edge of the pulse, where dI/dt is positive, the instantaneous frequency is down shifted, which is analogous to red shift, see Fig.9. Similarly for the trailing edge, dI/dt being negative, the frequency is up shifted or "blue shift". The pulse is linearly chirped, since its frequency changes continuously from leading edge to trailing edge. Furthermore the z dependence means the overall chirp depends on the fiber length. The above treatment has neglected group velocity dispersion, inclusion of which will either compress or broaden the pulse temporally.

Figure 10, shows experimentally observed spectra, for a nearly gaussian input pulse, at the output of a 99 meter long fiber (33). The spectra are labeled by the maximum phase shift which is related linearly to the peak power. The nonlinear phase shift increases with pulse intensity, labeled below each figure, and so does the pulse spectral width. In addition to the spectral broadening, another notable feature is that the SPM induced spectral broadening is accompanied by an oscillating structure covering the entire frequency range. This is in agreement with the theoretical calculations (33), which shows that the oscillating structure arises from interference between the same frequency components occurring in two different parts of the pulse. The chirped pulse can be exploited for pulse compression and generation of optical solitons.

Pulse Compression

Ultra short duration optical pulses are of scientific interest and have applications in such diverse fields as optical spectroscopy and generation of ultra short electrical pulses for semiconductor device characterization. Pulse compression can be accomplished by first chirping the pulse, followed by manipulating the velocity of the different frequency components. Note that chirping is essential, since the frequency spectrum of a pulse is inversely proportional to its temporal width. SPM is effective in impressing a linear chirp on a high intensity optical pulse. The linearly chirped pulse can be compressed if its trailing edge catches up with the leading edge during propagation. The basic principles of optical pulse compression can be understood with reference to Fig.11. In this case 532 nm wavelength, 33 picosecond long pulses are coupled in a 105-m long fiber. The fiber impresses a linear chirp, and due to the positive GVD, the pulse is temporally broadens as well. The chirped pulse is passed through a grating pair compressor, which essentially provides negative GVD. This can be inferred from Fig. 11, wherein the blue shifted spectrum has a shorter path length, between the grating pair, as compared to that for the red shifted spectrum. The double pass through the grating enhances this effect. In essence after passing through the grating pair, the trailing edge of the pulse catches up with the leading edge. Using this experimental set up, Johnson *et.al* (34) realized a pulse compression of 80 *i.e* the 33 picosecond pulse was compressed to 0.5 picosecond. In a similar experiment, but using single pass through the grating, 620 nm wavelength 40 femtosecond long pulses were compressed to 8 femtoseconds by Knox *et.al* (35). This compressed pulse represents less than 4 optical cycles! The ultra short pulses are of interest for such diverse applications as optical spectroscopy, and high bit rate optical communications.

Optical pulses at wavelengths exceeding 1.3 μm would experience both SPM and negative (anomalous) GVD during their propagation in silica fiber. Such a fiber can itself compress a pulse without the need of an external grating pair. The fiber compressor requires proper choice of input pulse intensity, pulse shape and fiber length, and is referred to as soliton-effect compressor, as described in the following.

Optical Solitons

The term solitons refers to special kinds of waves that can propagate undistorted over long distances and remain unaffected after collision with each other. Fundamental optical solitons are light pulses which maintain their shape and width with propagation distance. In this case the effect of nonlinearity (SPM), which broadens the spectral width of the pulse, is compensated by the negative group velocity dispersion (GVD) in the fiber (Fig. 2). It is apparent that for this to materialize the pulse intensity should be high enough to achieve the desired spectral broadening. Furthermore, the wavelength should be in the negative GVD region, which generally implies wavelengths longer than 1.3 um. In addition, a detailed analysis (36) shows that in the absence of loss, the input pulse should have hyperbolic secant shape. The peak pulse power for the fundamental soliton is given by (37),

$$P = \beta_2 / \gamma T_o^2$$

where β_2 and γ are parameters related to fiber GVD and SPM respectively, and T_o the width of the incident pulse (4). For power levels much above this, higher order solitons (38) are formed, and for power levels below this ordinary pulse propagation occurs.

Optical solitons are of considerable interest for high bit rate long distance communication systems. This is because, one of the factors that limits the bandwidth of an optical communication system is the pulse distortion that arises from dispersive spreading. The 1.5 μm wavelength region, where fiber loss is minimum and GVD is negative, is well suited for soliton based optical communications. Recently, using a 41.7 km fiber loop, a laboratory demonstration of soliton propagation for over 4000 km has been reported (39). In this experiment the fiber loss was compensated through Raman gain. What is most impressive about the experimental result is that even after 4000 km of propagation, the pulse retained its original shape.

Stimulated Inelastic Scattering

A light ray generally experiences scattering when it encounters any kind of inhomogenities of the medium in its path. For example the Rayleigh scattering is due to density and composition fluctuations at a scale smaller than the wavelength of the light. In this case, and for the third order nonlinear effects described so far, no energy exchange takes place between the medium and the electromagnetic wave. These are elastic scattering processes. The scattering processes accompanied by energy transfer between the medium and the propagating wave are called inelastic scattering. Two such processes, Raman and Brillouin scattering are of critical importance in optical communications. In spontaneous Raman scattering, light encountering the molecules, imparts part of its energy to the vibrational mode (optic mode) of the molecule. The scattered light, down shifted in frequency, is called the Stokes shifted component. Similarly, spontaneous Brillouin scattering is light scattered off the acoustic modes of vibrations. At high enough optical intensities, both of the above scattering processes become stimulated, so that the scattered light has laser-like characteristics. Furthermore, at high intensities, the excited molecules may impart part of their energy to the incident light, up shifting the frequency of the light. This scattered component is called anti-Stokes.

Stimulated Raman Scattering (SRS)

Spontaneous Raman scattering is light scattered off certain Raman active optic modes of vibration. The energy difference between the incident and the scattered radiation therefore represents the energy of the vibrational mode, *i.e*

$$\hbar(\omega_1-\omega_2) = \hbar\nu$$

where ω_1, ω_2, are incident and scattered radiation frequencies and ν the vibrational mode frequency. In crystalline materials the vibrational frequencies are spread out in discrete narrow bands. Due to inhomogeneous broadening, these discrete bands overlap in amorphous materials, resulting in a broad continuum. Figure 12 shows the Raman scattering cross section (40) of three principal glasses used in optical fibers, which reflect the spectral features of their respective molecular vibrations, with P_2O_5 providing the higher frequency shift. We note that GeO_2 and P_2O_5, which are used as index enhancing dopants in optical fibers, possess higher scattering cross section than silica. However, their concentration comprises less than 15% (molar) in low loss silica fiber. The Raman spectra of these fibers is therefore mainly that of silica. However, high germania and phosphorous doped fibers are of interest for high Raman gain (41) and large frequency shifts (42).

Spontaneous Raman scattering converts only a small fraction (typically 10^{-6}) of the incident light to Stokes shifted component. At high incident (pump) beam intensities , the scattering increases by several orders of magnitude, resulting in conversion of nearly all of the pump beam to the Stokes shifted component. As mentioned earlier, this is called Stimulated Raman Scattering, and was first reported by Woodbury and Ng (43).

For a continuous wave excitation, the evolution of the scattered (SRS) and the pump intensities along the fiber length can be described by the following two coupled equations.

$$dI_s/dz = g_r I_p I_s - \alpha_s I_s$$

$$dI_p/dz = -(\omega_s/\omega_p) g_r I_p I_s - \alpha_p I_p$$

where g_r is the Raman gain coefficient, and α_p, α_s, fiber loss coefficients for the scattered and the pump wavelengths respectively. The Raman gain coefficient g_r is related to the spontaneous Raman scattering cross section and can be determined experimentally. Figure 13 shows the measured (44) Raman gain cross section for silica fiber. The key feature to note here is that the gain peaks at 13.2 THz (440 cm^{-1}) and is of broad spectral width. It is easy to see that the above two coupled equations are based on energy conservation. The solution of the first of the above two equations, for $I_p >> I_s$, shows that the scattered beam grows exponentially along the fiber length.

To see how the process of SRS initiates, we note that a cw pump beam will result in Stoke shifted components spanning the frequency spectrum of the fiber Raman cross section. As the pump intensity is increases, the Stoke shifted component with the highest Raman gain will dominate and build up exponentially. For silica fibers this occurs at frequency shift of 440 cm^{-1}. However if a frequency selective feedback is provided, the gain at the feed back frequency will dominate. Both of these concepts are used in constructing devices based on SRS.

Fiber Raman laser, operating simultaneously at multiple Stokes frequencies (45), exploits both the gain peak in the Raman cross section and the feed back principle. Fig.14 is a schematic of one such laser. Mirrors M_1, M_2, M_3, provide feedback for first, second, and third Stokes components, and the grating is used to separate the Stokes components at the output. Since in low loss fibers, Raman gain scales with length, only a few watts of pump power is sufficient to achieve SRS. In fact the fiber Raman laser described in reference (45) used only 1 W of pump power at 514 nm, in a 10 meter long fiber. In this configuration the gain occurs for resonant cavity modes.

For higher pump powers stimulated gain for a single-pass can be attained. The first demonstration of single-pass Raman generation in the near-infrared in optical fibers was by Lin *et.al* (46). Using pulses of as high as 1 kW peak power at 1060 nm (Q-switched Nd:YAG laser), in a 170 meter long fiber, they simultaneously generated several orders of Stokes components, Fig. 15. However, in pulsed mode of operation, useful fiber length is limited due to the group-velocity difference between the pump and the Stokes component. The maximum gain results in a length within which sufficient overlap is retained between the pump and the scattered wave.

Stimulated Brillouin Scattering (SBS)

Brillouin scattering is light scattered off the acoustic modes of vibration. Similar to SRS, at high pump intensities the process becomes stimulated. The acoustic mode frequencies are several orders of magnitude lower than the optical frequencies. This implies several differences between the SRS and SBS waves. First, the Stokes shifted Brillouin wave differs from pump by only several GHz. The pump source should therefore be of sufficiently narrow linewidth to observe this effect. In silica fibers the Brillouin frequency shift is about 11.5 GHz. Second, the wave vector conservation allows only backward propagating SBS wave. The gain (47) for the SBS is two to three orders of magnitude higher than that for SRS. The back scattered SBS wave, therefore, can be quite intense, often containing more than 50% of the incident pump energy. The intense back scattered wave can generate an acoustic shock (48), causing damage to the fiber end. For narrow linewidth lasers the threshold for SBS can be as low as a few milli watts. In a communication system SBS causes attenuation of the signal in the forward direction and sends intense stoke-shifted light back towards the transmitter. These effects are much pronounced in coherent systems, since they use narrow linewidth lasers.

Summary

This paper has reviewed a few of the important nonlinear effects in optical fibers. The topics covered were selected to illustrate some of the key features of nonlinear effects in optical fibers, and hence is not a comprehensive review of the subject. Topics such as, modulation instability, four photon mixing, and photosensitivity are few other effects not discussed. The interested reader should consult the general references mentioned in the introductory section. Finally, the recent development of rare earth doped (50) silica fibers should result in new avenues of research for nonlinear fiber optics. This is because, rare earth doped fibers combine nonlinear effects and optical gain in the same device.

Acknowledgments

The author thanks Chinlon Lin, Yaron Silberberg, and Andy Weiner for many interesting discussions. He is specially indebted to Roger Stolen of AT&T Bell Labs for helpful discussions and for several figures acknowledged in the text.

References

1. R. H. Stolen, Chapter 5 in "Optical Fiber Telecommunications" Editors S.E. Miller, A.G. Chynoweth, Academic Press (1979)
2. Section IV, Optical Phenomena in Fibers, " Physics of Fiber Optics", Editors B. Bendow , S. S. Mitra, The Am. Ceram. Soc., Inc., (1981)
3. Chinlon Lin, IEEE J. Lightwave Technology, LT-4, 1103(1986)
4. G.P. Agrawal, Nonlinear Fiber Optics, Academic Press(1989)
5. Y. Fuji, B.S. Kawasaki, K.O. Hill and D.C. Johnson, Opt. Lett. 5, 48(1980)
6. Y. Sasaki and Y. Ohmori, Appl. Phys. Lett. 39, 166(1981)
7. A. S. L. Gomes, U. Osterberg, and J. R. Taylor, Appl. Phys.B 41, 235(1986)
8. U. Osterberg and W. Margulis in Technical Digest of the Fourteenth IQEC,p. 102.
9. U. Osterberg and W. Margulis, Opt. Lett. 11, 516(1986)
10. R. W. Terhune and D. A. Weinberger J. Opt. Soc. Am. B 4, 661(1987)
11. R. H. Stolen and H. W. K. Tom, Opt. Lett. 12, 585(1987)
12. N. Bloembergen, Nonlinear Optics, Benjamin, New York(1965)
13. M. C. Farries, P. St.J. Russel, M. E. Fermann and D. N. Payne Electron. Lett. 23,322(1987)
14. V. Mizrahi, U. Osterberg, J. E. Sipe and G. I. Stegeman, Opt. Lett. 13, 279(1988)
15. M. E. Fermann, M.C Farries, P. St. J. Russel and D. N. Payne, Opt. Lett. 13, 282(1988)
16. U. Osterberg and W. Margulis, Opt. Lett. 12, 57(1987)
17. M. A. Saifi and M. J. Andrejco, Opt. Lett. 13, 773(1988)
18. T. E. Tsai, M. A. Saifi, E. J. Friebele, D. L. Griscom and U. Osterberg, Opt. Lett. 14, 1023(1989)
19. S. R. Friberg, Y. Silberberg, M. K. Oliver, M. J. Andrejco, M. A. Saifi, and P. w. Smith, Appl. Phys. Lett. 51, 1135(1987)
20. S. M. Jensen, IEEE J. Quantum Electron. QE-18, 1580(1982)
21. A. W. Snyder and Y. Chen, Opt. Lett. 14, 517(1989)
22. Y. Chen, A. W. Snyder, and D. J. Mitchell, Electron. Lett. 26, (77)1990
23. V. Mizrahi, K. W. DeLong, G. I. Stegeman, M. A. Saifi, and M. J. Andrejco, Opt. Lett. 14, 1140(1989)

24. S. J. Aitchison, M. K. Oliver, E. Kapon, E. Colas and P. W. E. Smith, Appl. Phys. Lett., in press
25. G. I. Stegeman and R. H. Stolen, J. Opt. Soc. Am. B 6, 652(1989)
26. S. R. Friberg, A. M. Weiner, Y. Silberberg, B. G. Sfez and P. W. Smith, Opt. Lett. 13, 904(1988)
27. A. M. Weiner, Y. Silberberg, H. Fouckhardt, D. E. Leaird, M. A. Saifi, M. J. Andrejco and P. W. Smith, IEEE J. Quantum Electron. 25, 2648(1989)
28. N. J. Doran and D. Wood, J. Opt. Soc. Am. B 4, 1843(1987)
29. R. H. Stolen and A. Ashkin, Appl. Phys. Lett. 22, 294(1973)
30. S. Rashleigh, J. Light Wave Technol. LT-1, 312(1983)
31. G. P. Agrawal, Chapter 4, Nonlinear Fiber Optics, Academic Press (1989)
32. E. P. Ippen, C. V. Shank and T. K. Gustafson, Appl. Phys. Lett. 24, 190(1974)
33. R. H. Stolen and C. Lin, Phys. Rev. A 1448(1978)
34. W. H. Knox, R. L. Fork, M. C. Downer, R. H. Stolen, C. V. Shank and J.A.Valdmanis, Appl. Phys. Lett. 46, 1120(1985)
35. A. M. Johnson, R. H. Stolen, and W. M. Simpson Appl. Phys. Lett. 44, 729(1984)
36. A. Hasegawa and F. Tappert, Appl. Phys. Lett 23, 142 (1973)
37. L.F. Mollenauer, R. H. Stolen and J. P. Gordon, Phys. Rev. Lett. 45, 1095 (1980)
38. N. J. Doran and K. J. Blow, IEEE J. Quantum Electron. QE-19, 1883 (1983)
39. L. F. Mollenauer and K. Smith, Opt. Lett. 13, 675(1988)
40. F. L. Galeener, J. C. Mikkelsen , R. H. Geils and W. J. Mosby, Appl. Phys. Lett. 32, 34 (1978)
41. H. Takahashi, J. Chang, K. Nakamura, I. Sugimoto, A. Oyobe and Y. Fuji, Opt. Lett. 11, 383 (1986)
42. V. V. Grigor'yants *et.al* , Opt. Quant. Electron. 9, 351(1977)
43. E. J. Woodbury and W. K. Ng, Proc. IRE 50, 2347 (1963)
44. R. H. Stolen, C. Lee and R. K. Jain, J. Opt. Soc. Am. B1, 652 (1984)
45. R. H. Stolen, C. Lin, J. Shah, and R. F. Leheny, IEEE J. Quantum Electron. QE-14, 860 (1978)
46. L. G. Cohen and C. Lin, IEEE J. Quantum Electron. QE-14, 855 (1978)
47. G. A. Koepf, D. M. Kalen and K. H. Greene, Electron. Lett. 18, 942 (1982)
48. Y. Azuma, N. Shibata, T. Horiguchi, M. Tateda, Electron. Lett. 24, 250 (1988)
49. D. Cotter, J. Opt. Commun. 4, 10 (1983)
50. P. Urquhart, IEE Proceedings ,Pt. J., 135, 385(1988)

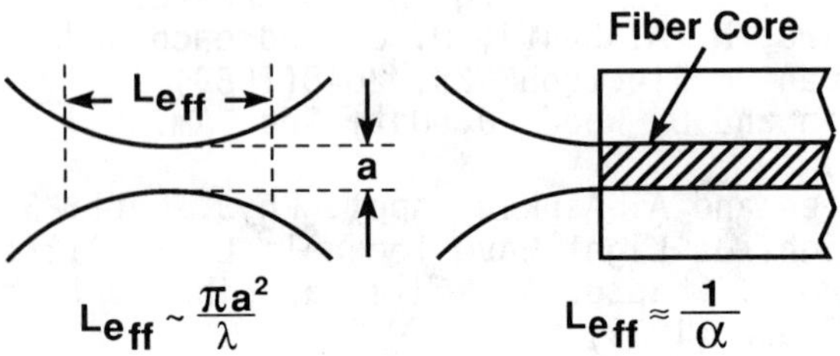

- **High Field Intensity**
- **Long Interaction Length**
- **Phase Matching**

Fig. 1 Key Features

Nonlinear Optics in Fibers

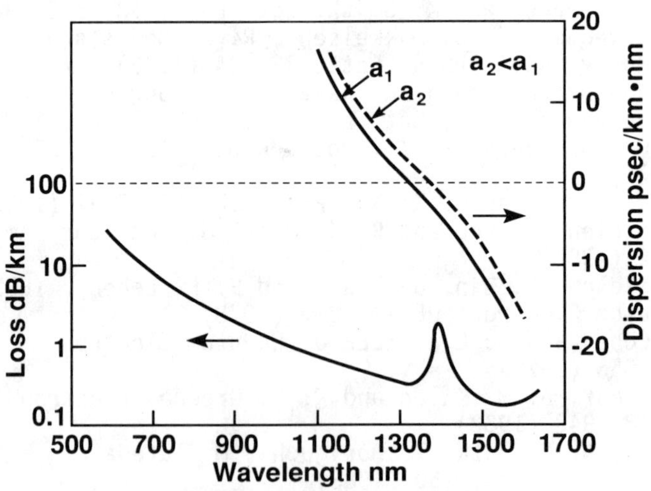

Fig. 2 Fiber Loss and Dispersion

Note: a_1,a_2 are core diameters.

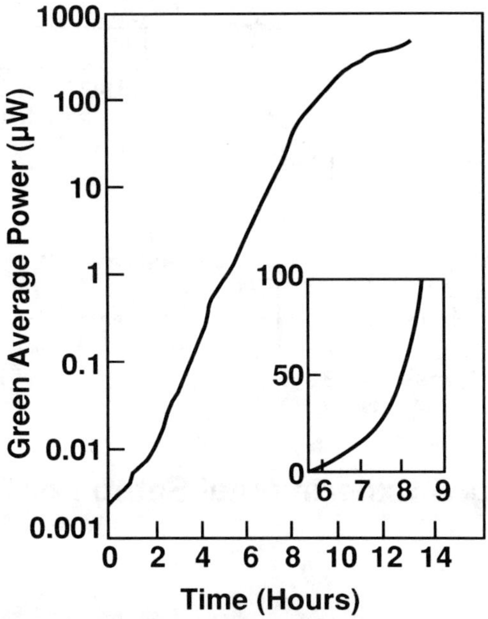

Fig. 3 Efficient SHG In Glass Fibers

From Reference (9) Osterberg et. al.

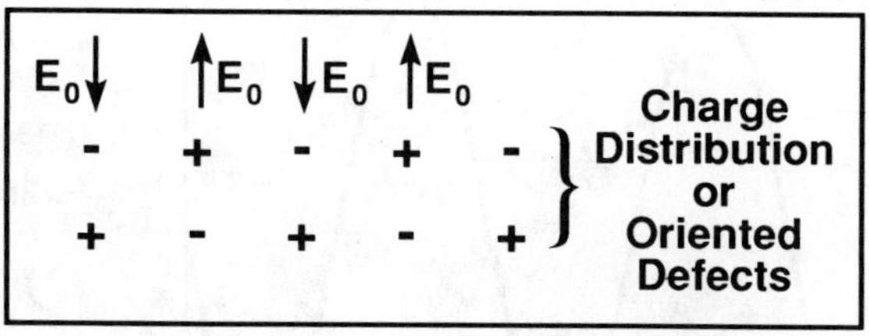

$$E_0 = \chi^{(3)} E_1^* E_1^* E_2 \exp i\,(k_2 - 2k_1)Z$$

Fig. 4 Model For $\chi^{(2)}$ Grating

Based on Reference (11) Stolen and Tom

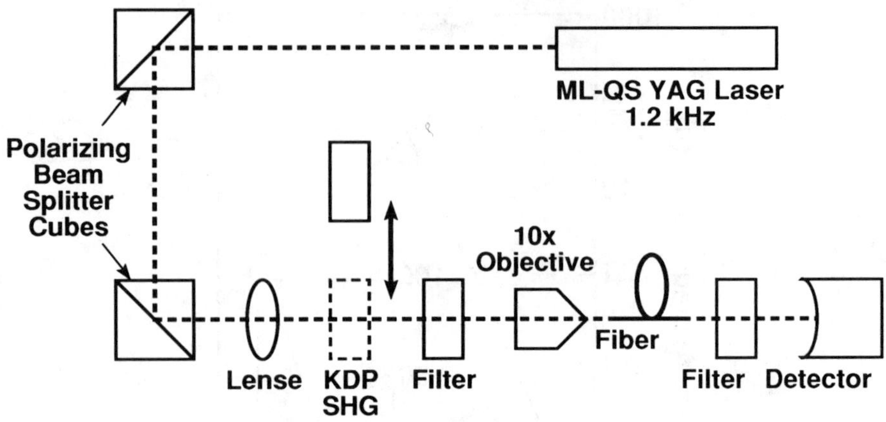

Fig. 5 Experimental Setup For SHG

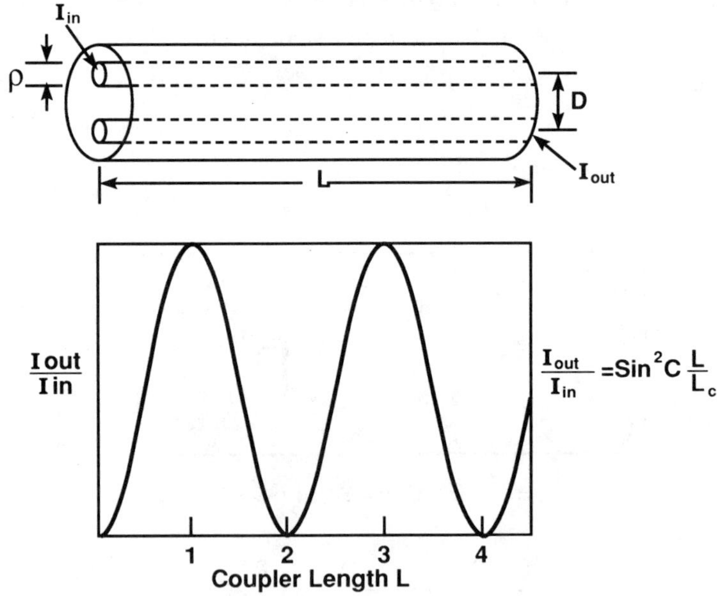

Fig. 6 Directional Coupler

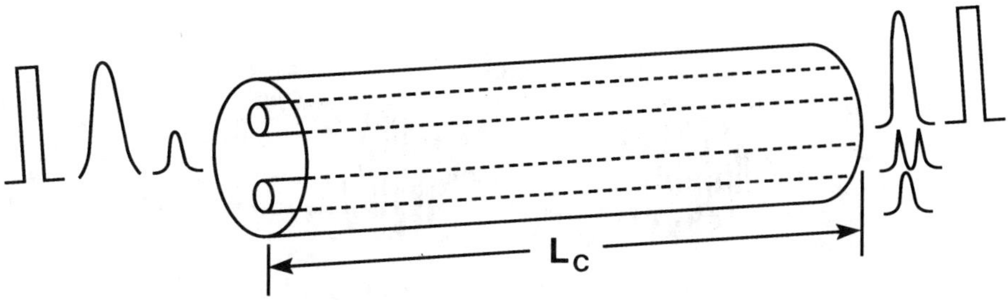

Fig. 7 Switching In Nonlinear Directional Coupler

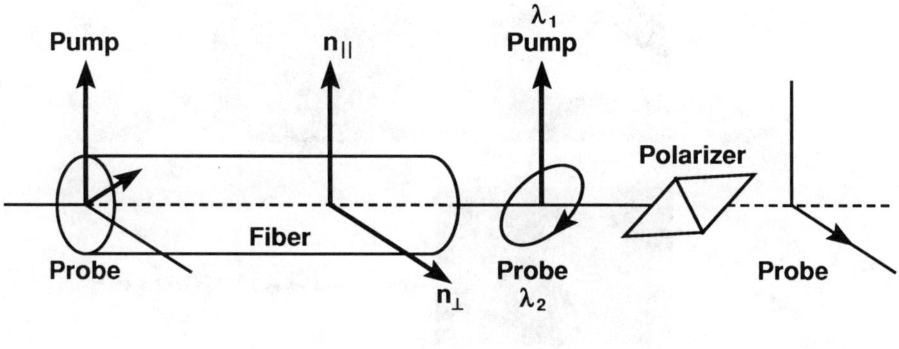

Fig. 8 Optical Kerr Effect
After Reference (29) Stolen and Ashkin

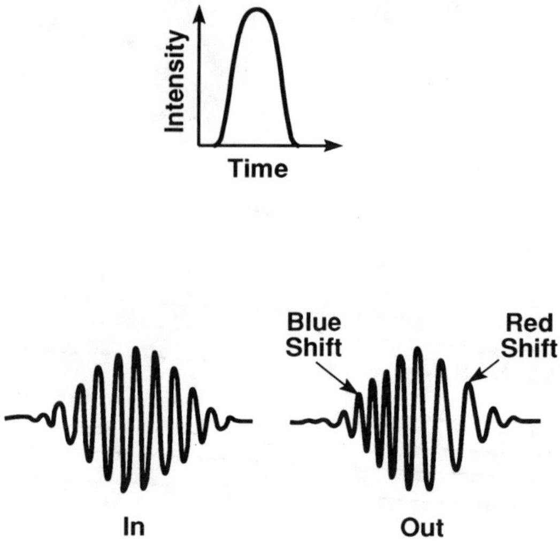

Fig. 9 Effect of Self-Phase Modulation on the Pulse Instantaneous Frequency

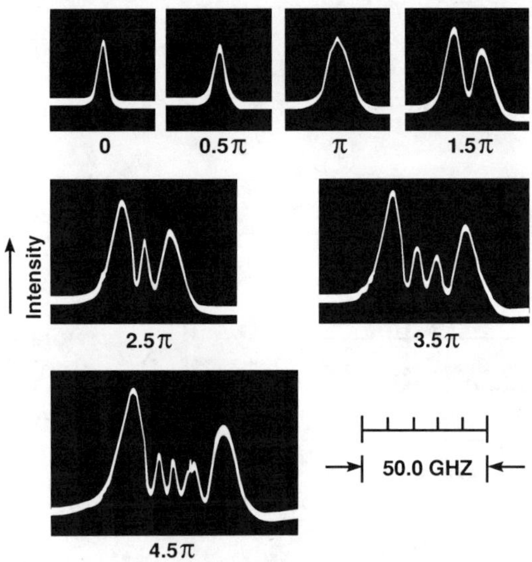

Fig. 10 Frequency Spectrum

After Reference (33) Stolen and Lin

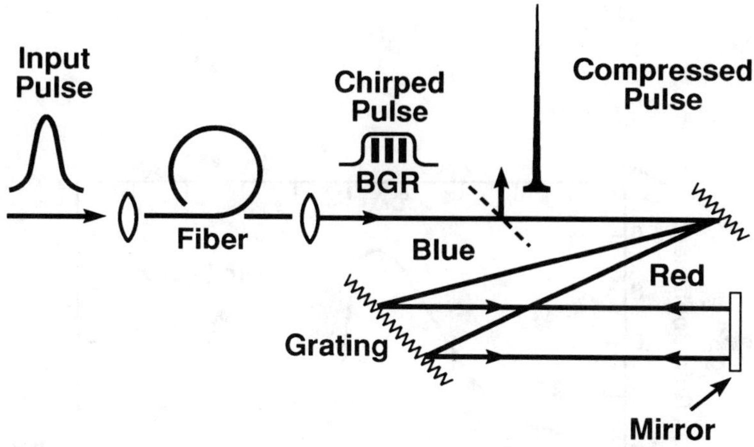

Fig. 11 Pulse Compression Using Fiber and Grating Pair

From reference (35) Johnson, et. al.

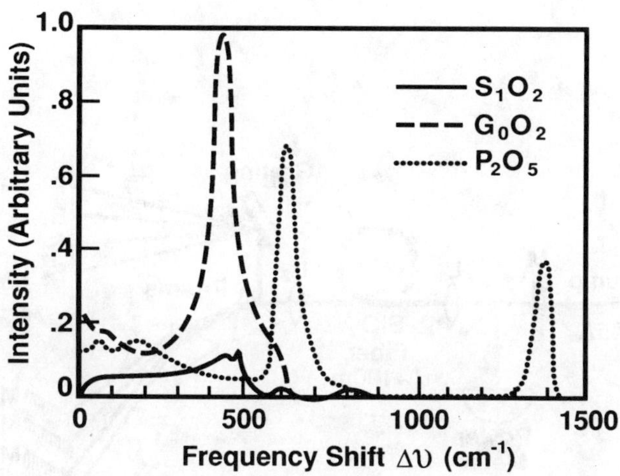

Fig. 12 Spontaneous Raman Scattering Spectra of Some Oxide Glasses

After Reference (40) Galeener, et. al.

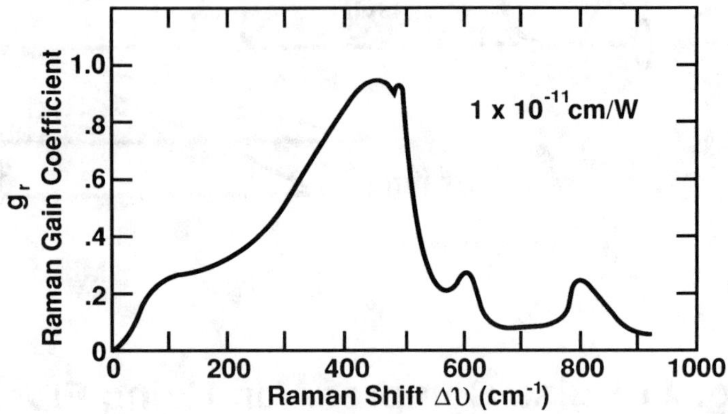

Fig. 13 Raman Gain Spectra of a Typical Silica Glass Fiber
After Reference (44)

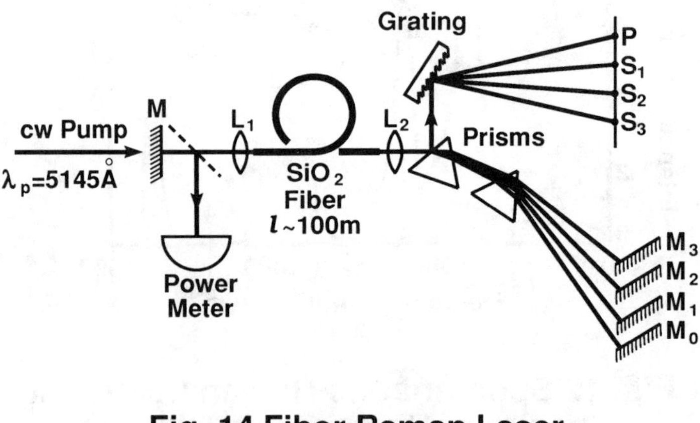

Fig. 14 Fiber Raman Laser
After Reference (45) Stolen et. al.

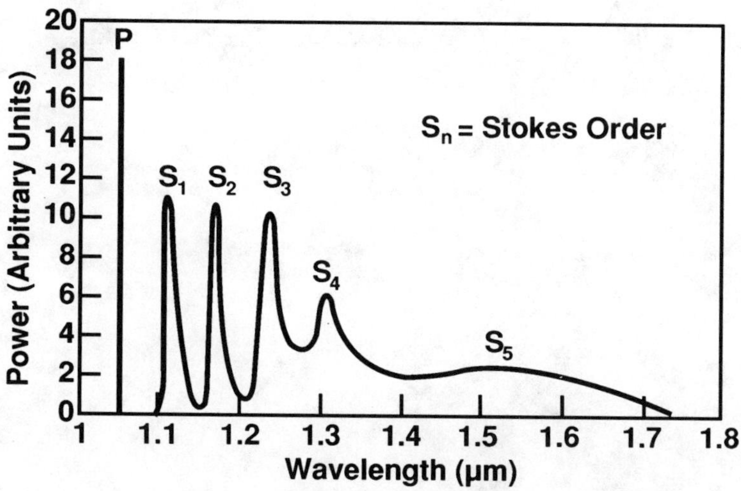

Fig. 15 Typical Multiple Stokes Spectra Generated By Single-pass Formulated Raman ScatteringIn A Single Mode Silica Glass Fiber

After Reference (46) Cohen and Lin

DEFECT CENTERS INDUCED BY THE HARMONIC WAVELENGTH OF 1.06 μm LIGHT IN Ge-P-DOPED SILICA

T. E. Tsai
GEO Centers, Inc., Fort Washington, MD 20744

E. J. Friebele and D. L. Griscom
Naval Research Laboratory, Washington, DC 20375

M. Saifi
Bell Communication Research, Red Bank, NJ 07701

ABSTRACT

To investigate the generation mechanism of the Ge E′ centers associated with second harmonic generation (SHG) in the Ge-P-doped silica core fibers, electron spin resonance (ESR) was used to study the defect centers induced by the third and fourth harmonic wavelengths of 1.06 μm light in Ge-P-doped silica preform. We have found that fourth harmonic light readily induced the Ge E′ center, while third harmonic light could induce the center only at high power. Ge(1), Ge(2) and phosphorus oxygen hole (P-OHC) centers were also observed. The fact that only Ge E′ center is observed in SHG fibers can be explained in terms of the radiation and thermal bleaching of these centers during SHG treatment.

I. INTRODUCTION

We have reported[1] recently that the second harmonic generation efficiency in Ge-and Ge-P-doped silica-core single-mode fibers is correlated with the light-induced Ge E′ center concentration. Understanding the generation mechanism(s) of Ge E′ center is therefore not only interesting scientifically but also important technologically.

To induced high-efficiency (> 1 %) SHG in Ge- and Ge-P-doped silica-core single mode fibers, injection of high-power (≈ 40 to 10 kW peak power) pulses from a mode-locked Q-switched Nd:YAG laser with or without a green seeded light for a period of time is necessary[2]. At such high laser power, harmonic wavelengths can be induced through nonlinear interactions of the injected light with core material[3]. It is then interesting and important to know which harmonic wavelength of the injected light induced the Ge E′ centers.

In the irradiation of Ge- and Ge-P-doped silica core fiber with X- or -rays, several types of Ge-associated centers were reported[4]. Detailed studies[5,6,7] suggested that in general they can be associated with trapped electron centers (Ge(1) and Ge(2)) and trapped hole center (Ge E′). Ge E′ and Ge(1) were induced by UV-light with 92% of spectral output at 254 nm[8,9], while Ge E′ and Ge(2) were induced by UV-light of a broad spectral output varying from 2 to 5 eV[10]. In addition, with P as a co-dopant SHG conversion efficiency as high as 13% can be induced in Ge-doped silica core fibers[11]. It is then important to know the contribution, if any, of these Ge(1), Ge(2) and P-associated centers in the induced SHG effect.

In this paper, ESR studies of defect centers induced by the third and fourth harmonic wavelengths of 1.06 μm light in Ge-P-doped silica preform will be reported. We have found that fourth harmonic light readily induced the Ge E′, P-OHC[12], Ge(1) and Ge(2) centers, while third harmonic light could induced the Ge E′ and P-OHC centers only at high power. The fact that only the Ge E′ center is observed in SHG fibers will be discussed in terms of their radiation response and thermal stability.

II. EXPERIMENTAL

Samples of size 2mm x 2mm x 10mm were cut from Ge-P-doped silica preforms. At room temperature, they were irradiated with unfocused beam of a multigas excimer laser (model EMG 103 MSC, Lambda Physik) with either KrF or XeF as the lasing medium (corresponding approximately the fourth and third harmonic wavelength, respectively, of 1.06 μm IR light). The power of the laser was measured with an excimer power meter (model EPM 300, Lambda Physik). X-band ESR spectra ($\approx$ 9.4 GHz) were obtained with a Bruker model ER 200D-SRC spectrometer employing 100 kHz field modulation. The sample temperature during the ESR measurements was maintained at 120 K by means of a modified Bruker nitrogen flow-through device.

III. RESULTS AND DISCUSSIONS

A. Defect Centers Induced by 3.5 eV Photons (Third Harmonic Wavelength)

No defect can be induced in Ge-P-doped silica by irradiation with 3.5 eV photons (generated by a XeF excimer laser) of pulse energy

10 mJ/cm^2 and pulse width 20 ns for a period of 50 minutes (The observed Ge E′ centers are the pre-existing ones). However, as shown in Fig. 1, Ge E′ centers can be induced when pulse energy of 80 mJ/cm^2 is used. At a fluence of 8 x 10^7 mJ/cm^2, P-OHC is also observed, as seen in Fig. 2. (Note, the P-OHC is known[12] to have a second hyperfine peak which in the present case is overlapped and obscurred by the Ge E′ center.)

B. Defect Centers Induced by 5.0 eV Photons (Fourth Harmonic Wavelength)

Ge E′, Ge(1), Ge(2) and P-OHC centers, and non-bridging oxygen hole centers (NBOHC) can be induced readily by irradiation of Ge-P-doped silica with 5.0 eV photons (generated by a KrF excimer laser) of pulse energy 10 mJ/cm^2 and pulse width 20 ns. The ESR spectrum at a fluence of 1.2 x 10^4 mJ/cm^2 (2 minutes of irradiation at 10 Hz) is shown in Fig. 3(a). The spectrum of the Ge-associated centers was computer simulated with concentration ratios of Ge E′:Ge(1):Ge(2)=1:10:10. Clearly at this low fluences the dominant Ge-associated centers are Ge(1) and Ge(2) centers.

The ESR spectrum of the sample irradiated to a fluence of 2.3 x 10^8 mJ/cm^2 is shown in Fig. 3(b). Ge E′, Ge(1), Ge(2), NBOHC, P-OHC centers and peroxy radicals ($\equiv Si\text{-}O_2^\cdot$) are observed. The concentration ratios of the Ge-associated centers, obtained by computer simulation of the observed spectrum, are Ge E′: Ge(1): Ge(2) = 56 : 1 : 1. At this fluence, the dominant Ge-associated defect centers are Ge E′ centers.

C. Radiation Response of the Induced Defect Centers

The radiation response of the observed paramagnetic centers as a function of the fluence of 5.0 eV photons is shown in Fig. 4. We find that the concentrations of Ge E′, P-OHC and peroxy radicals increase with fluence, while those of Ge(1), Ge(2), and NBOHC centers quickly reach maximum at low fluence, and then decrease by about a factor of 2 over the next 3 decades of fluence. The latter centers drop of sharply at the highest fluence employed.

Since the decrease in the concentration of NBOHC is correlated with the increase in the concentration of peroxy radicals, radiation-induced reaction of NBOHCs ($\equiv Si\text{-}O^\cdot$) with interstitial oxygen to form $\equiv Si\text{-}O_2^\cdot$ is likely to have occurred (thermally induced conversion of NBOHCs to peroxy radicals was also observed in the irradiated high purity silica. Details of this phenomenon will be discussed in future publication).

Even though the decreases of Ge(1) and Ge(2) centers are

accompanied with the growth of Ge E′ center at fluences greater than 10^4 mJ/cm^2, these decreases and the growth are not one for one. Thus, there is no quantitative correlation of the decrease in the concentration of Ge(1) and Ge(2) centers with other defect centers observed, and the decrease of Ge(1) and Ge(2) centers with irradiation fluence is likely the result of radiation-bleaching. The bleaching of Ge(1) and Ge(2) centers by a broadband xenon arc lamp (the shortest wavelength component is about 5.0 eV) was also reported in Ref. 8.

Ge E′ is the only Ge-associated center induced in the irradiation of Ge-P-doped with 3.5 eV photons. Its radiation response is shown in Fig. 1.

D. Thermal Stability of the Induced Defect Centers

In the studies of SHG fibers using ESR, it is important to know the thermal stability of the induced defect centers at room temperature (isothermal annealing) since SHG fibers are kept at room temperature before ESR measurements is carried out. The detailed studies of the thermal stability of the induced Ge-associated centers in Ge-doped silica in isochronal annealing experiments was reported by Friebele, et. al.[13] and that of P-associated centers in P-doped silica reported by Griscom et. al.[12]. These data indicate that most of the observed defect centers reported here are (meta)stable at room temperature.

To determine the stability of the observed defect centers at room temperature, samples in which defect centers were induced at the highest fluences using 3.5 and 5.0 eV photons shown in Figs. 1 and 4 were kept at room temperature and the defect concentrations were monitored by ESR over a year. No significant change in Ge E′ and P-OHC centers concentrations were observed. However, a decrease in peroxy radical and an increase in NBOHC populations were noted (see Fig. 3 (b) and (c)). Similar behavior can also be observed in Fig. 4 at fluence about 3 x 10^7 mJ/cm^2 where ESR measurements were carried out right after the irradiation and after storing the sample at room temperature overnight.

E. Implications on the Generation Mechanism of Ge E′ Center Under SHG Preparation Process

In the SHG preparation process, high power (1 to 40 kW peak power) IR light at wavelength 1.06 μm is injected into a single-mode Ge-doped silica core fiber of core diameter $\approx$ 5 μm (the corresponding power density is 3 x 10^9 to 10^{11} W/cm^2). Using the third order susceptibility of silica fibers[14] of the order of 10^{-22} m^2/V^2, the maximum power density of the third harmonic wavelength induced by

40 kW pulses is less than 250 W/cm^2. Since no defect can be induced by the third harmonic wavelength of power density ≈ 5 x 10^5 W/cm^2 (for a pulse energy 10 mJ/cm^2 and pulse width 20 ns) as shown in Fig. 1, Ge E′ centers are then not induced by the third harmonic wavelength of the 1.06 μm light during SHG preparation processes.

On the other hand, Fig. 1 suggests that no threshold power density is required for the generation of Ge E′ centers with the fourth harmonic wavelength. Since silica is a centrosymmetric material, no even order susceptibilities are anticipated. However, weak even order susceptibilities can exist in silica fiber at the core-cladding interface and can also result from quadruple and magnetic dipole moments of silica. These can initially induce (very weak) fourth harmonic wavelength for the generation of Ge E′ centers in SHG fibers prepared without a green seeded light. The very weak fourth harmonic wavelength induced initially in the preparation processes is consistent with the long preparation time (≈ hours) required in the non-seeded SHG preparation process.

In the seeded SHG preparation processes, in-phase second harmonic wavelength (2) of 100 W peak power is injected with 1 kW peak power of 1.06 μm light ( ). In this case, the fourth harmonic wavelength can be induced either by mixing 2 , and through third order susceptibility or by mixing 2 and 2 through the second order susceptibility. The intensity of the fourth harmonic wavelength induced in this way is much higher than that induced in the non-seeded processes (about 50 times that induced in non-seeded case if both are induced by third order susceptibility). The short preparation time (≈ minutes) required in this seeded preparation processes can then be anticipated.

F. The Absence of Ge(1), Ge(2) and P-OHC Centers in SHG Fibers

As mentioned previously, Ge(1), Ge(2) and P-OHC centers can also be readily induced by the fourth harmonic wavelength (≈ 5.0 eV photons). However, none of these centers are observed in SHG fibers[1]. The absence of Ge(1) and Ge(2) centers in SHG fiber can be explained in term of the photo-bleaching of these centers by photons with energies equal or less than 5.0 eV discussed above.

The absence of P-OHC center in the Ge-P-doped silica core SHG fibers can not be understood in terms of its radiation response since its population increases (albeit slowly) with irradiation fluence of both third and fourth harmonic wavelengths (see Fig. 4). Nor can it be interpreted by thermal bleaching at room temperature since P-OHCs are not thermal bleached at room temperature (see Sect. D). More studies are needed to understand the absence of this center in Ge-P-doped SHG fibers.

REFERENCES

1. T. E. Tsai, M. A. Saifi, E. J. Friebele, D. L. Griscom, and U. Osterberg, Opt. Lett. 14, 1023 (1989).

2. W. Margulis and U. Osterberg, J. Opt. Soc. Am. B5, 312 (1988).

3. G. P. Agrawal, "Nonlinear Fiber Optics", Academic Press, Inc. New York (1989), p. 14-15.

4. E. J. Friebele, D. L. Griscom and H. Sigel, Jr., J. Appl. Phys. 45, 3424 (1974).

5. H. Kawazoe, J. Non-Cryst. Solids, 71, 231 (1985).

6. T. E. Tsai, D. L. Griscom, E. J. Friebele, and J. W. Fleming, J. Appl. Phys. 62, 2264 (1987).

7. T. E. Tsai, D. L. Griscom and E. J. Friebele, Diffusion and Defect Data 53-54, 469 (1987).

8. (a) R. N. Schwartz, G. L. Tangonan, G. R. Blair, W. Chamultrat and L. Kevan, "Electron Paramagnetic Resonance and Optical Study of Radiation-Induced Defect Centers in Doped Silica Glasses", in Defects in Glasses, Eds. by F. L. Galeener, D. L. Griscom and M. J. Weber (Materials Research Society, Pittsburg, 1986), p. 197.

9. R. N. Schartz and G. R. Blair, J. Appl. Phys. 65, 710 (1989).

10. J. Simpson, J, Rirger, and F. Dimarcello, "UV-Radiation Induced Color Centers in Optical Fibers", ibid. 8, p. 333.

11. M. C. Farries, M. E. Fermann and P. St. J. Russell, "Second Harmonic Generation in Optical Fibers", in Nonlinear Guided-Wave Phenomena: Physics and Application, 1989 Technical Digest Series, Vol. 2 (1989), p. 246.

12. D. L. Griscom, E. J. Friebele, K. L. Long and J. W. Fleming, J. Appl. Phys. 54, 3743 (1983).

13. E. J. Friebele and D. L. Griscom, "Color Centers in Glass Optical Fiber Waveguides", ibid. 8, p.319.

14. ibid. 3, p. 19.

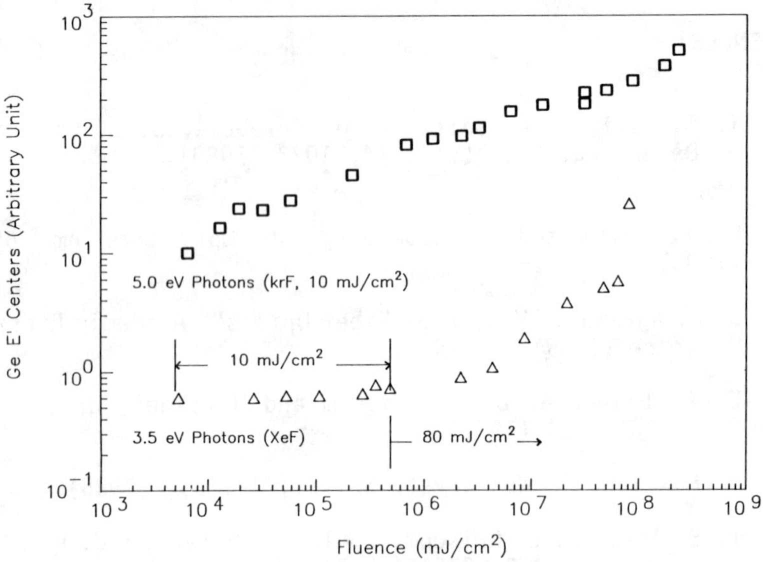

1. The concentration of Ge E′ centers in a Ge-P-doped silica as a function of the fluence of 5.0 eV and 3.5 eV photons.

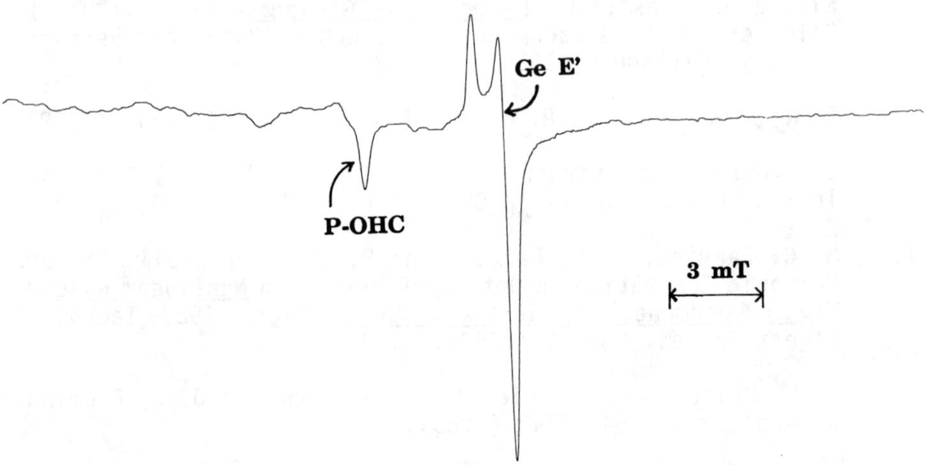

2. 120 K ESR spectrum of Ge-P-doped silica irradiated with 3.5 eV photons of fluence s 8×10^7 mJ/cm².

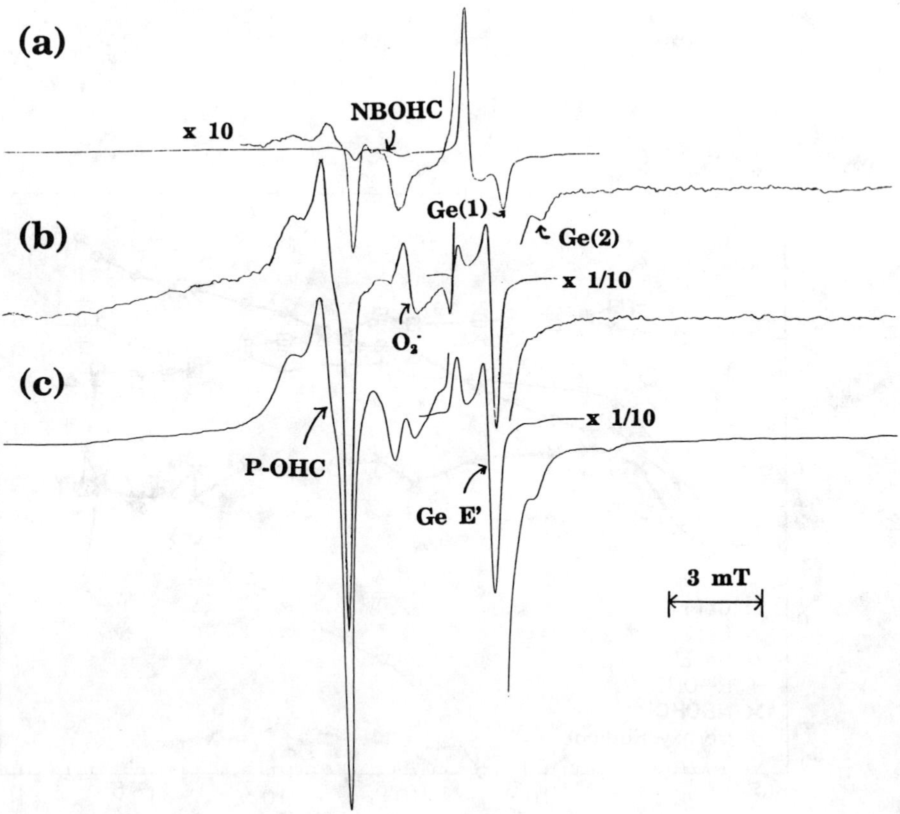

3. 120 K ESR spectrum of Ge-P-doped silica irradiated with 5.0 eV photons (a) of fluence 1.2 x 10^4 mJ/cm^2, (b) of fluence 2.3 x 10^8 mJ/cm^2, and (c) of the fluence of (b) but measured one year after irradiation.

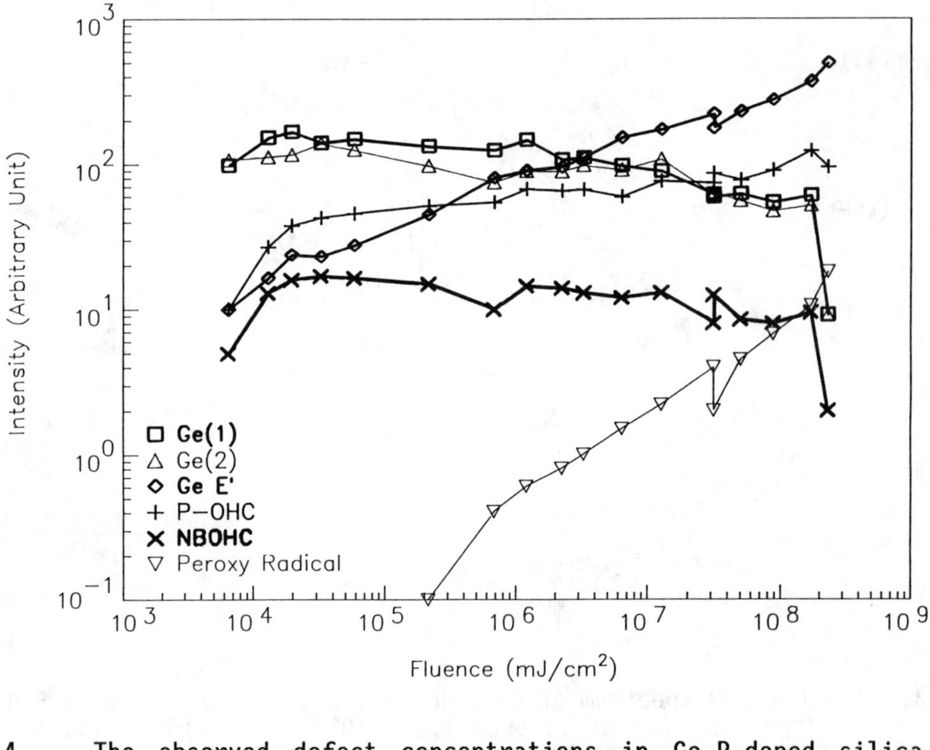

4. The observed defect concentrations in Ge-P-doped silica induced by 5.0 eV photons as a function of fluence.

THIRD-ORDER NONLINEAR OPTICAL EFFECTS IN POLYANILINE DOPED SILICA FILMS VIA THE SOL-GEL TECHNIQUE

E.T. Knobbe, B. Dunn, P.D. Fuqua, F. Nishida, R.B. Kaner* and B.M. Pierce†, Department of Materials Science and Engineering and *Department of Chemistry, University of California, Los Angeles and †Hughes Aircraft Company, Long Beach California.

ABSTRACT

Doped oxide and mixed oxide materials are promising candidates for the development of nonlinear optical devices. Conducting polymers have been shown to possess large $\chi^{(3)}$ values, but generally exhibit poor processibility. The sol-gel synthesis technique provides a method by which soluble organic species can be incorporated into rigid oxide matrices. A model system has been investigated wherein a silica gel host was doped with 2-ethyl polyaniline (2-Et PANi). Films containing 9.6×10^{-4} M 2-Et PANi were synthesized and found to have third-order susceptibilities of 4.8×10^{-13} esu, via degenerate four-wave mixing experiments.

INTRODUCTION

Conjugated polymers are good candidates for third-order nonlinear optical materials due to their extended polarizable network of delocalized π-electrons. Many, including polyacetylene and polyaniline (PANi), have been shown to possess large resonant and nonresonant third-order susceptibilities, $\chi^{(3)}$, in bulk or solution form [1], [2]. Conducting polymers in general, however, exhibit poor processibility. Thus, attempts to fabricate devices using conducting polymers, such as electro-optic modulators in the form of waveguide films, have been slowed by the somewhat intractable nature of these conjugated organic molecules.

Sol-gel matrices can provide an inert host capable of supporting optically active guest materials [3]-[5]. In prior work we have demonstrated that the sol-gel process can be used to fabricate solid-state tunable dye lasers [6], [7]. The sol-gel solution technique offers a low temperature synthesis route whereby soluble organics can be doped into porous oxide matrices. The resulting host material exhibits good optical properties while maintaining chemical and thermal stability. Melt glass processing, by comparison, generally requires temperatures in excess of 500° C, greatly exceeding the decomposition temperatures of most organic molecules.

The sol-gel technique has been described by numerous investigators, and involves the hydrolysis and condensation of organometallic precursors such as tetramethoxy silane (TMOS) [8], [9]. Condensed oxide particles coalesce into a porous gel network, entrapping compatible organic molecules such as the recently developed soluble conducting polymer materials.

For the purpose of this study, silica gels were doped with 2-ethyl polyaniline, a derivative form of PANi. The third-order susceptibility of doped gels was determined via the degenerate four-wave mixing technique. Results are compared to a carbon disulfide (CS_2) reference solution and to undoped silica gel films.

BACKGROUND

$\chi^{(3)}$ and Degenerate Four-Wave Mixing

Under the influence of applied electric fields (E), the intrinsic electronic susceptibility of a material is not constant, but varies as a function of local field amplitude. The total susceptibility can be expressed as a power series expansion such that:

$$\chi(E) = \chi^{(1)} + \chi^{(2)}E + \chi^{(3)}EE + \ldots \quad . \tag{1}$$

The local polarization is a function of χ and E such that:

$$P(\chi, E) = \chi(E)E \ . \tag{2}$$

In the case of small or moderate field amplitudes, $\chi(E)$ converges to the first order susceptibility $\chi^{(1)}$ and the resultant polarization is linear with respect to E. When the internal fields become large, however, nonlinear polarization terms are introduced through the higher order susceptibilities. The total polarization, P^T can be expressed as the sum of the linear (P^L) and nonlinear (P^{NL}) polarization terms as [10], [11]:

$$P^T(\chi, E) = P^L(\chi, E) + P^{NL}(\chi, E) \ , \tag{3}$$

where:

$$P^L(\chi, E) = \chi^{(1)}E \tag{4}$$

and $$P^{NL}(\chi, E) = \chi^{(2)}EE + \chi^{(3)}EEE + \ldots \quad . \tag{5}$$

The linear relationship in expression (4) is associated with material characteristics such as refractive index, optical absorption or gain and birefringence. These properties are related to classical optics wherein electromagnetic waves pass through each other without influencing one another by mixing. In noncentrosymmetric materials, the second order term in (5) gives rise to effects such as second harmonic generation, parametric mixing, optical rectification, and Pockels effect. The cubic term relates $\chi^{(3)}$ to four-wave mixing phenomena which include frequency tripling, stimulated Raman and Brillouin scattering, dc Kerr effect, and degenerate four-wave mixing. Due to symmetry considerations, purely amorphous materials exhibit only odd ordered optical nonlinearities, while anisotropically oriented media may possess nonzero values for both even and odd ordered susceptibilities.

For the purpose of this study, degenerate four-wave mixing (DFWM), also referred to as the instantaneous ac Kerr effect or optical phase conjugation, was used to determine the third-order nonlinear susceptibility of doped silica gel films. The term "phase conjugation" is used because the output signal wave propagates as the time reverse or phase conjugate of the input probe beam. The phase conjugate mirror is a grating which is formed in the presence of two pump waves and a probe wave within the nonlinear medium. Quantitatively, the propagating electromagnetic waves can be described by the following:

$$\mathbf{E}_i(\mathbf{r}_i,t) = \frac{1}{2}\left\{\mathbf{A}_i(\mathbf{r}_i)\exp\left[j(\mathbf{k}\bullet\mathbf{r}_i - \omega t)\right] + \text{C.C.}\right\}, \tag{6}$$

where **E** is the electric field, **A** is the field amplitude and **r** is the distance travelled in the direction of **k**, the wave vector. C.C. indicates complex conjugate. The wave vector has a magnitude k described by:

$$k = \frac{\omega n}{c}, \tag{7}$$

where c is the speed of light in vacuum and n is the refractive index of the material at the free-space wavelength (having frequency ω) of interest.

Figures 1a and 1b demonstrate the generation of volume phase gratings which are constructed in DFWM experiments. Standing waves are produced as a result of the interaction between E_1, E_2, and E_3. The fields E_1 and E_2 are defined to be counterpropagating phase fronts; therefore $\mathbf{k}_1 = -\mathbf{k}_2$. The superposition of these standing waves creates a periodically modulated polarization grating, according to equation 3. This spatially modulated polarization field results in a volume phase grating in these materials, which can be decomposed into two coupled holograms; a reflection grating (figure 1a) and a transmission grating (figure 1b) [10]. The result is diffraction of a portion of the incident field intensities into the phase conjugate wave, E_4. The four-wave mixing condition is satisfied when:

$$\mathbf{k}_1\omega_1 + \mathbf{k}_2\omega_2 + \mathbf{k}_3\omega_3 + \mathbf{k}_4\omega_4 = 0 \tag{8}$$

Since the frequencies are degenerate and the pump waves counterpropagate, $\mathbf{k}_3$ must be equal to $-\mathbf{k}_4$. The phase conjugate signal wave propagates along the $\mathbf{k}_4$ wave vector direction. The magnitude of the coupled output wave is dependent upon a number of factors including the incident wave amplitudes, losses due to inelastic scattering events and the third-order susceptibility.

Caro and Gower have examined the phenomenon of phase conjugation in absorptive media [12]. Their work included the effects of depleted pump/probe field amplitudes, which have been shown to have a direct impact upon P^{NL}, as in expression (5). Following the work of Caro and Gower, Dennis, Blau and Bradley reported optical phase conjugation in polydiacetylene [13] polymer systems. Borrowing from this earlier research, the phase conjugate reflectivity, R_{PC}, is defined to be to be the ratio of the phase conjugate intensity I_4 with the probe beam intensity I_3:

$$R_{PC} = \left(\frac{A_4(0)}{A_3(0)}\right)^2 = \left(\frac{I_4}{I_3}\right) . \tag{9}$$

Assuming the DFWM geometry shown schematically in figure 2, $\chi^{(3)}$ is related to R_{PC} by [12]:

$$\chi^{(3)} = \frac{4c^2n^2\varepsilon_o\alpha\sqrt{R_{PC}}}{3\omega T(1-T)I_1} , \tag{10}$$

where ε_o is the permittivity of free space, α is the optical absorption coefficient and I_1 is the incident pump wave intensity. T is the sample transmissivity such that:

$$T = \exp[-\alpha \ell] , \tag{11}$$

$$\alpha = \varepsilon N . \tag{12}$$

where ℓ is the interaction length, ε is the extinction coefficient and N is the active species number density. The third order susceptibility is linearly dependent upon N as:

$$\chi^{(3)} = \gamma N L_3 \cdot \tag{13}$$

γ is the orientationally averaged third-order molecular polarizability coefficient and L_3 describes local field effects for third order processes. Equation (10) assumes the low phase conjugate reflectivity case as described in [12]. Implicit in this treatment is the assumption that multi-photon absorption effects are negligible.

Conducting Polymers and Polyaniline

Conjugated polymer systems have been the focus of considerable interest for more than ten years due to their high conductivity. This behavior arises from an extensively conjugated polymer backbone. Conductivity is developed in these polymers upon addition of suitable dopant species, and can be continuously varied from the insulating to the semiconducting or metallic forms by utilizing dopants of the proper electroactivity and concentration. Electrical conductivity has been demonstrated in a large number of conjugated polymer systems, including polyacetylene, polydiacetylene, polyparaphenylene, polypyrrole, polythiophene, polyaniline, and various "ladder" polymers [14].

As described in the Introduction, many of the conducting polymers, including PANi, have been shown to possess large third-order susceptibilities. Polyaniline is unusual among the conducting polymer family in that it exhibits solubility in various organic solvents. PANi, for example, can be partially dissolved into dimethyl formamide (DMF) and N-methyl pyrrolidinone (NMP). Modification by suitable substituents, such as short chain alkylation, can make the polymer (mol wt $\geq$ 5000) completely soluble in a variety of solvents, even producing modest solubility in alcoholic solutions [15].

PANi can be converted from its partially oxidized, intrinsically insulating emeraldine base (EB) form to the conducting emeraldine salt (ES) form by protonation via aqueous acid solution [16]. The emeraldine base material is the partially soluble polymer form, and appears as a violet-blue compound in solution. The conducting form is relatively intractable, usually appearing as a green precipitate. Incorporation of the soluble emeraldine base polymer into sol-gel matrices raises the interesting prospect of *in-situ* doping of the PANi within the porous oxide matrix. For the purpose of this study, 2-ethyl polyaniline (2-Et PANi) in the partially oxidized emeraldine base form was incorporated into a rigid silica gel host by the sol-gel solution method. The structure of EB 2-Et PANi is given in figure 3. Measurements of the third-order susceptibility were performed via the degenerate four-wave mixing technique. The $\chi^{(3)}$ values for doped gel films were calculated, and compared to those of undoped gel films and to a CS_2 control sample.

EXPERIMENTAL METHODS

Sample Preparation

2-ethyl polyaniline, with reported molecular weight of 5000, was prepared by the method outlined by Leclerc *et al.* [15]. As described in the previous section, the soluble free amine form (EB) of PANi is known to be readily protonated under acidic conditions, producing the highly insoluble ammonium salt form (ES) [16], [17]. In order to maintain the free amine form in solution, it was necessary to synthesize the silica gel in the presence of a minimal amount of acid catalyst.

The initial silica sol consisted of 11 ml of tetramethoxy silane (TMOS), 5 ml of distilled water, and 1 drop of 12 N HCl. A monophasic solution was prepared by the "sonogel" method, after Esquivas and Zarzycki [18]. Under continuous sonication, a total of 30 ml of tetraethoxy silane (TEOS), 7.5 mL of distilled H_2O, and 11 ml of TMOS were added in a dropwise fashion to form the precursor solution. The final silica sol was composed of a 1 : 1 : 5 mole ratio of TEOS : TMOS : H_2O. The acid catalyst was present at a level of 1×10^{-3} : 1 molar ratio of acid to alkoxide. A minimal volume of water was used in the reaction in order to form a more suitable solvent for the 2-Et PANi.

Doping of 2-Et PANi into the silica sol was accomplished by dissolving the polymer into a mutually compatible solvent. N-methyl pyrrolidinone (NMP) was chosen as the carrier solvent due to the good solubility of 2-Et PANi in NMP and its compatibility with the silanol solution. NMP containing approximately 2.5 weight percent 2-Et PANi was added to a silica sol-NMP mixture. The final composition of the solution was 7 : 1 : 2 volume ratio of silica sol : NMP : 2-Et PANi/NMP. After the addition of all components, the solution was filtered through a 0.7 μm glass fiber filter and subsequently placed into styrene containers for gelation. Two weeks after the onset of gelation the closed containers were opened, and residual solvent was allowed to evaporate. The film samples were dried in air at room temperature for three weeks prior to optical analysis.

Optical Characterization

The magnitude of the third-order susceptibility, $\chi^{(3)}$, was determined by the degenerate four-wave mixing method [11]. The pump/probe wave source was an actively Q-switched Nd:YAG laser. All measurements were performed at 1.06 μm, with Q-switched pulses of 15-20 nsec (FWHM). Nominal power output of the laser was 5 MW. Source waves were focussed to an area of approximately 8 mm^2 at the sample plane, as in figure 2. Pump/probe wave intensities (I_1, I_2, and I_3)

were approximately matched, magnitudes of 2.5 $\times 10^{11}$ W/m^2 each at the sample plane ("S" polarized). The Nd:YAG laser did not contain an etalon device, resulting in the generation of relatively short coherence length (≤20 mm) pulses. For this reason, L_1, L_2, and L_3 were path length matched to within 10 mm. The counterpropagating phase conjugate wave, I_4, was detected by means of a partial reflection from the beamsplitter BS_2. The signal was detected using a reverse biased *p-i-n* silicon photodiode, which transmitted current pulses to a Tektronix 2430 digital storage oscilloscope. Measurements were conducted on thick (1 mm) film samples, with the I_1 and I_2 waves intercepting the samples near normal incidence. The angle between I_1 and I_3 was determined to be 7 degrees. To eliminate any lensing effects due to slight warping of the dried gel films, all samples (doped and undoped) were held within a liquid gate filled with an index matching fluid (mineral oil). The CS_2 reference liquid was measured in a 1 mm path length silica glass cuvette. Background signal amplitudes from the liquid gate and silica glass cuvette were measured and subtracted from the observed gel and reference liquid sample response values.

RESULTS AND DISCUSSION

The 2-ethyl polyaniline concentration in the silica gel film was determined by constructing a Beer's law calibration curve from solutions of known concentration. Assuming an average molecular weight of 5000, the 2-Et PANi concentration in the silica gel was found to be 9.6 $\times 10^{-4}$ M. The refractive indices of CS_2 and 2-Et PANi:SiO_2 were estimated to be 1.6 and 1.4 at 1.06 µm, respectively. The emeraldine base doped silica gel was found to have low losses due to scatter, and exhibited good transparency at 1.06 µm. Spectrophotometric measurements at 1.06 µm yielded absorption coefficients of 0.1 cm^{-1} (> 99% T over 1 mm pathlength) for the CS_2 reference and 0.4 cm^{-1} (96% T over 1 mm pathlength) for the 2-Et PANi doped silica film. The phase conjugate reflectivity, R_{PC}, of the doped gel was found to be 31% of that from the CS_2 reference. Using expressions (10)-(13), and assuming the third-order susceptibility of carbon disulfide to be 1.7 $\times 10^{-12}$ esu at 1.06 µm [19], $\chi^{(3)}$ of the polyaniline doped gel was calculated to be 4.8 $\times 10^{-13}$ esu (6.7 x 10^{-21} m^2/V^2). Peak response of the doped gel was found to be delayed by approximately 3-4 nsec with respect to CS_2. Based upon the prior work of others, the observed third-order susceptibility of the doped gel is thought to be due to exciton formation, corresponding to electronic transitions in PANi as opposed to other possible polarization phenomena such as electrostriction [20], [21]. Although the magnitude of this nonlinearity is not exceedingly large, it is very encouraging to observe significant nonlinear behavior in a model system for which the host matrix has been shown to be highly processible. It is believed that the third order susceptibility of the doped film can be greatly improved by raising the concentration of 2-Et PANi in the host gel, as described later.

Material processibility and optical quality are extremely important for device considerations. The fabrication of waveguide films, for example, can result in devices having very large nonlinear effects [22]. Assuming that the damage threshold of the material is sufficiently high, it is relatively easy to achieve enormous (>1 GW/cm^2) pump/probe field intensities within a thin film structure. According to equations (5), (9) and (10), the phase conjugate output signal is closely tied to the local field amplitudes. Confinement of the waves in this fashion, therefore, would be expected to significantly increase the resultant output signal.

When examining optical phase conjugation behavior in absorptive media, it is of interest to evaluate the changes in output signal response as a function of variables such as the dopant level. By inspection of (10)-(13), it can be seen that phase conjugate reflectivity is a function which is dependent upon a number of interrelated parameters. Although $\chi^{(3)}$ increases linearly with number density N, absorption losses due to large concentrations (increased α) result in significantly reduced pump/probe field intensities. Phase conjugate reflectivity, therefore, should exhibit saturation behavior as a function of increased 2-Et PANi levels. Using the previously computed values for third-order nonlinear susceptibility and absorption coefficient, a theoretical plot showing the change in normalized R_{PC} as a function of 2-Et PANi dopant concentration is given in figure 4. The actual 2-Et PANi concentration used in this study is indicated (+). Assuming a 1 mm thick film, the peak theoretical reflectivity occurs at an 2-Et PANi concentration of approximately 1.2 $x10^{-2}$ M, representing a twelve-fold increase in the currently obtainable dopant concentration. This dopant level corresponds to an increase of the phase conjugate reflectivity by a factor of 22. The expected $\chi^{(3)}$ of the film at that concentration was computed to be approximately 5 $x10^{-12}$ esu, roughly three times larger than that of pure carbon disulfide. The comparable output signal for a nonabsorptive material, CS_2, is shown in figure 5. The relative concentration used in the DFWM experiments is indicated (+). Clearly it would be of great interest to improve the solubility of polyaniline in the precursor solution. It is also important to note, however, that achieving significantly greater solubility than this value is of limited value. Only in cases where the absorption losses approach zero can extremely high phase conjugate reflectivities, assuming moderate field intensities over long interaction lengths, be obtained.

SUMMARY AND FURTHER STUDY

The third-order nonlinear optical properties of a model system, comprised of a derivative form of polyaniline entrapped in a silica gel host, have been explored. Preliminary results indicate that the films possess $\chi^{(3)}$ values of 4.8 $\times 10^{-13}$ esu at 1.06 μm. Because doped silica gel films may be processed into the form of optical devices, this composite system is of interest for further investigation. Continued study will be required in order to determine whether dopant concentrations can be increased to a level such that doped oxide gels become viable alternatives to other nonlinear materials systems.

ACKNOWLEDGEMENTS

The authors gratefully acknowledge the contributions to this project made by Mr. Eun Woo Chang, Mr. Ben Mattes, and Ms. Deanna Chen. This research has been supported by the National Science Foundation (DMR 87-06010, ENG 88-11629 and CHE 86-57822) and by the Lockheed Missiles and Space Company, Palo Alto, CA.

BIBLIOGRAPHY

[1] Nonlinear Optical Properties of Organic Molecules and Crystals, Vols. 1 and 2, J. Zyss and D.S. Chemla eds., Academic Press, 1988.

[2] S. Ermer, K.P. Aron, G.A. Hansen, G.F. Lipscomb, R. Lytel, J.I. Thakara, A.J. Ticknor and S.I. Yaniger, "Conjugated Polymers as Third-Order NLO Materials, Proc. of SAMPE, June, 1989 (to be published).

[3] B. Dunn, E. Knobbe, J.M. McKiernan, J.C. Pouxviel, and J.I. Zink, "The Optical Behavior of Organic and Organometallic Molecules in Sol-Gel Matrices", MRS Symp. Proc., C.J. Brinker, D.E. Clark, D.R. Ulrich eds., MRS Society, Pittsburgh, **121** (1989) 331-42.

[4] Y. Kobayashi, Y. Kurokawa and Imai, "A Transparent Alumina Film Doped with Laser Dye and its Emission Properties", J. Non-Cryst. Solids, **105** (1988) 198-200.

[5] C.A. Capozzi and L.D. Pye, "Preparation and Characterization of PMMA Modified SiO2 Host Material for Organic Dye Molecules", Proc. SPIE, **970** (1988) 135-142.

[6] E.T. Knobbe, B. Dunn, P.D. Fuqua, F. Nishida and J.I. Zink, "Laser Behavior in Organic Dye Doped Sol-Gel Materials", Proc. Fourth Intnl. Conf. Ultrastructure Processing, J. Non-Cryst. Solids, in press.

[7] E.T. Knobbe, B. Dunn, P.D. Fuqua and F. Nishida, "Laser Behavior and Photostability Characteristics of Organic Dye Doped Silicate Gel Materials", Applied Optics, in press.

[8] Proc. Third Intnl. Conf. on Ultrastructure Processing, J.D. Mackenzie and D.R. Ulrich eds., Wiley (1988).

[9] "Better Ceramics Through Chemistry", C.J. Brinker, D.E. Clark and D.R. Ulrich eds., Mat. Res. Soc. Proc., **73** (1986).

[10] Y. R. Shen, The Principles of Nonlinear Optics, First Edition, John Wiley & Sons, Inc., New York (1984).

[11] Optical Phase Conjugation, First Edition, edited by R. A. Fisher, Academic Press, Inc., New York (1983).

[12] R. G. Caro and M.C. Gower, "Phase Conjugation by Degenerate Four-Wave Mixing in Absorbing Media", IEEE J. Quantum Electron., **QE-18** (1982) 1376-1380.

[13] W. M. Dennis, W. Blau, and D. J. Bradley, "Optical Phase Conjugation in a Soluble Polymer", Optical Engineering, **25** (1986) 538-540.

[14] Proc. of the Third Intnl. Conf. on Sci. and Tech. of Synth. Metals, M. Aldissi ed., Synth. Metals., **27, 28, 29** (1989).

[15] M. Leclerc, J. Guay, and L.H. Ho, "Synthesis and Characterization of Poly(alkylanilines)", Macromolecules, **22** (1989) 649-653.

[16] A.G. MacDiarmid, J.C. Chiang, M. Halpern, W.S. Huang, S.L. Mu, N.L.D. Somasiri, W. Wu, and S.I. Yaniger, "'Polyaniline': Interconversion of Metallic and Insulating Forms", Mol. Cryst. Liq. Cryst., **121** (1985) 173-191.

[17] R.J. Cushman, P.M. McManus, and S.C. Yang, "Spectroelectrochemical Study of Polyaniline: The Construction of a pH-Potential Phase Diagram", J. Electroanal. Chem. , **291** (1986) 335-346.

[18] L. Esquivias and J. Zarzycki, Proceedings of the Third International Conference on Ultrastructure Processing, John Wiley & Sons Inc., New York (1988) 255-270.

[19] A. Yariv, Quantum Electronics, Third Ed., Wiley, New York (1988).

[20] M.G. Roe, J.M. Ginder, P.E. Wigen, A.J. Epstein, M. Angelopoulos and A.G. MacDiarmid, "Photoexcitation of Polarons and Molecular Excitons in Emeraldine Base", Phys. Rev. Lett., **60** (1988) 2789-2792.

[21] J.C. Altman, P.J. Elizando, G.F. Lipscomb, and R. Lytel, "Picosecond Nonlinear Optics of Organic Materials", Mol. Cryst. Liq. Cryst. Inc. Nonlin. Opt., **157** (1988) 515-523.

[22] G.I. Stegeman, C.T. Seaton and R. Zanoni, "Organic Films in Non-Linear Integrated Optics Structures", Thin Solid Films, **152** (1987) 231-263.

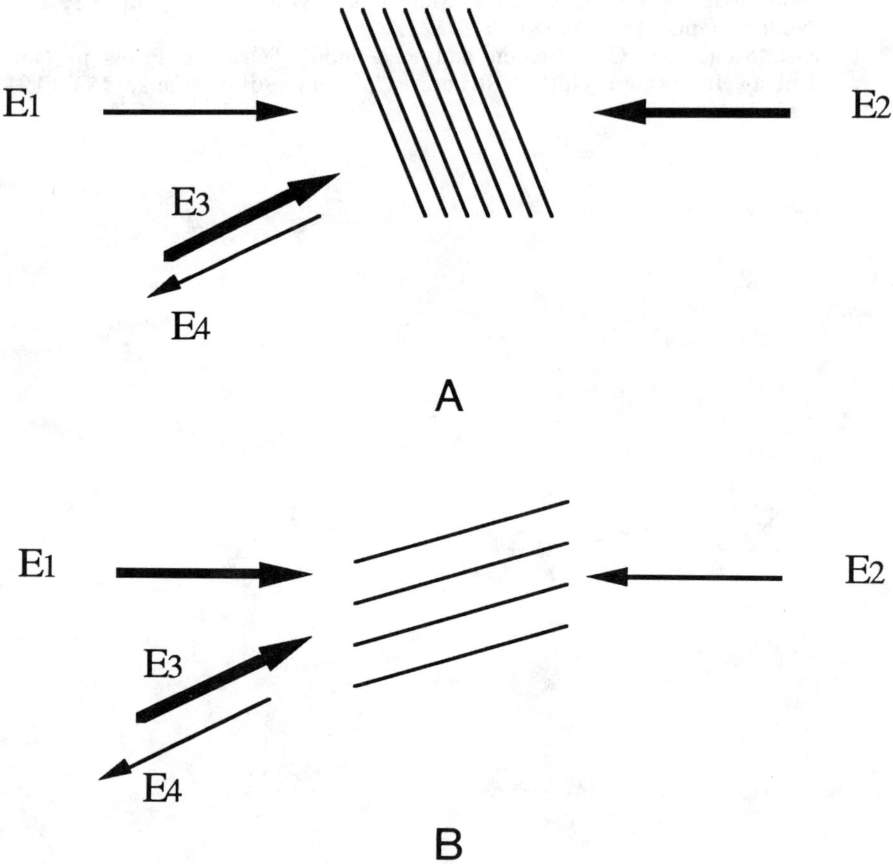

Figure 1. Reflection (a) and transmission (b) grating formation in phase conjugate mirror hologram.

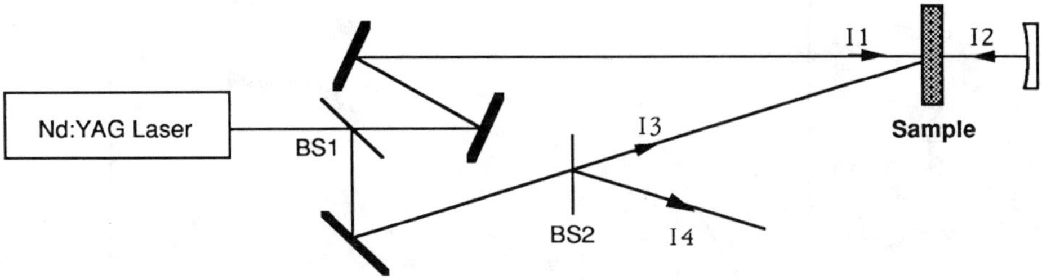

Figure 2. Schematic of degenerate four-wave mixing (DFWM) experiment. I_1, I_2, and I_3 are the pump/probe wave intensities while BS_1 and BS_2 indicate beam splitter locations.

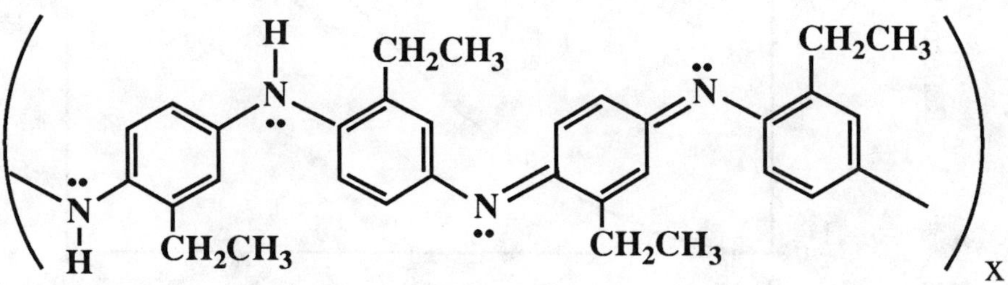

Figure 3. Idealized chemical structure of 2-ethyl polyaniline, emeraldine base form.

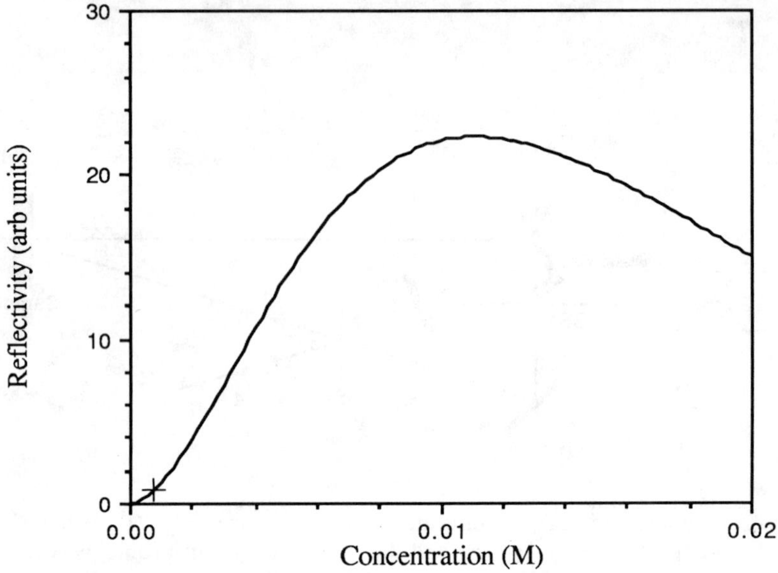

Figure 4. Phase conjugate reflectivity as a function of concentration in an absorptive medium (2-Et PANi:SiO_2).

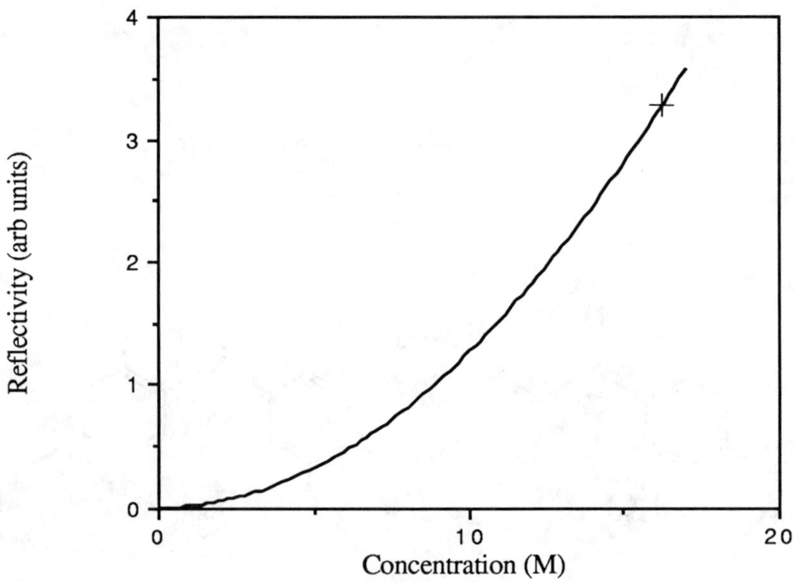

Figure 5. Phase Conjugate Reflectivity as a function of concentration in a nonabsorptive medium (CS_2).

OPTICAL STUDIES OF QUANTUM CONFINED CuCl- and $CuCl_xBr_{1-x}$GLASS COMPOSITES

J.A. Ruller,* R.J. Ginther*, G.M. Williams+, B.L. Justus,
A.J. Campillo, and E.J. Friebele
Naval Research Laboratory, Washington, DC 20375

ABSTRACT

Quantum confinement effects of semiconductor microcrystallites (quantum dots) doped in glasses are of current interest due to their potential for high optical nonlinearities. We report the precipitation of CuCl and CuClBr microcrystals in sodium aluminoborosilicate glasses. Control of the heat treatment permits the selective growth of quantum dots over a size range 50-125Å. We report preliminary measurements of the intensity-dependent absorption spectrum.

I. INTRODUCTION

Recently, interest in the optical properties of small semiconductor crystals dispersed in a glassy phase has increased because they exhibit large optical nonlinearities. It is known that as the crystal size decreases, the translational motion of the excitations of the semiconductor, whether electrons, holes, or excitons, becomes restricted, causing an increase in the minimum excitation energy. This in turn causes a blue shift of the absorption spectra. The size of the crystals is governed by the temperature and time of heat-treatment.

Much emphasis has been placed on $CdSe_xS_{1-x}$ and CdS microcrystals in silicate glasses,[1] developed for sharp cutoff filter applications. The $CdSe_xS_{1-x}$ system has proven to be complicated because of the stoichiometric changes that occur with heat-treatment time.[2,3] The optical properties of CuCl in a glassy matrix have also been investigated.[4-7] The CdS system has an exciton radius of 30Å, whereas the CuCl system has an exciton radius of about 7Å. The effects of the exciton screening on the properties of the CuCl system are being investigated in this study.

*Sachs Freeman Assoc., Inc., Landover, MD 20785
+VPI, Blacksburg, Va 24061

II. EXPERIMENTAL PROCEDURE

The copper halide glass presently being used is a photochromic glass composition purchased from Corning Glass Works. There are two very similar compositions of a sodium aluminoborosilicate glass that are being studied. One contains ~1% CuCl and the other contains ~1% $CuCl_xBr_{1-x}$. The glasses were melted in 1 kg batches for four hours at 1450°C. As melted, the glass is green because of the presence of both Cu^{+1} and Cu^{+2}, the former causing a yellow color and the latter giving a blue color to the glass.[8] The glass slabs were cut into 1 cm cubes and heat-treated at various temperatures/times in a Fisher model 495A ashing furnace. Samples that were used for optical measurements were cut to 100 μm and polished. Glass samples used for TEM and X-ray diffraction measurements were ground into fine powders.

The samples were examined using a JEOL 200CX Transmission Electron Microscope (TEM) in order to identify the crystals and determine the crystal size distribution. Figure 1 shows a glass that was heat-treated at 585°C for one hour. The microcrystallites appear as black spheres of about 100Å ± 20Å

To determine the crystalline phase in the glass containing only CuCl, X-ray diffraction data were obtained on the sample that was heat-treated for 4 hours at 585°. A Philips diffractometer with CuKα radiation was used. Because the microcrystals are so small, very long counting times were required. Figure 2 compares the measured X-ray diffraction with that of powdered CuCl. The good agreement between the two indicates that the phase is CuCl as expected.

Absorption spectra were taken at room temperature using a Varian Cary 2400. In Figure 3 the absorption spectrum of an unheat-treated CuCl sample is compared to that of a sample heat-treated for 30 minutes. An absorption peak at 368 nm and a shoulder at 377 nm appears upon heat-treating the glass for 30 minutes.

The absorption spectrum obtained at room temperature of a sample containing both chloride and bromide is shown in Figure 4. The peak is shifted to 383 nm and the shoulder is at 394 nm. The absorption spectrum of the same sample in a continuous flow cryostat at 77K is also shown in Figure 4. Two peaks are now clearly resolved and shifted to 388 nm and 399 nm, respectively.

Preliminary single pulse bleaching data were obtained at room temperature by probing 100 μm thick samples with the output of a traveling wave dye amplifier pumped by the frequency-tripled output of a Nd-glass laser. The broadband output was centered at 386 nm with a FWHM at 4 nm and pulse width of 90 psec.

III. DISCUSSION

Quantum confinement effects are expected in copper halide glasses when the size of the microcrystallites are on the order of the exciton radius. Previous theoretical treatments[9,10] have predicted large optical nonlinearities which increase with the size of the particles up to ~50Å diameter. Such large nonlinearities have been observed experimentally[11] for CuCl microcrystals in a NaCl host. In this paper we report preliminary studies of the growth and characterization of copper halide microcrystals of ~100Å diameter in glass. The TEM results indicate that the microcrystals in the sample heat-treated for 1 hour at 585°C (Figure 1) have a narrow distribution of sizes centered at ~100Å in diameter. The X-ray diffraction data confirm the chemical composition and zincblende structure of the microcrystals. The optical absorption spectra are in agreement with the spectra obtained by Ekimov, et al,[4-7] of bulk CuCl quantum dot glasses. Comparison of our spectra with the spectrum of bulk CuCl crystals indicates that the features we observe are most likely due to the spin-orbit splitting of the $Z_{1,2}$ and Z_3 exciton peaks. The absorptions at short wavelengths is attributed to a Cu^{+2} charge transfer band (230 nm) and Cu^{+1} (250 nm).[12,13] The single pulse bleaching results obtained at room temperature on the 388 nm resonance indicate that the nonlinear coefficients n_2 for the $CuCl_xBr_{1-x}$ glass is $\sim 10^{-12} m^2/w$. This value is in good agreement with previous measurements of n_2 in other semiconductor-doped glasses. We are actively pursuing further work on these systems at low temperatures.

This preliminary study has indicated the potential of CuCl and $CuCl_xBr_{1-x}$ glasses for nonlinear optical applications. However, to tailor and optimize their properties, it is necessary to more fully understand the crystal growth mechanisms and the effect of the glass composition on the crystallite formation. These studies are in progress, together with a complete characterization of the nonlinear optical behavior of these glasses.

The authors are grateful to David Morse of Corning Glass Works for valuable discussions. We would also like to thank Henry Hoff of NRL for his help with the TEM.

REFERENCES

1. D.W. Hall, N.F. Borelli, J. Opt. Soc. Am. Vol.5, No.8, 1650-1654, 1988

2. B.G. Potter,Jr., J.H. Simmons, Physical Review B, Vol. 37, No.18 (1987)

3. N.F. Borrelli, D.W. Hall, H.J. Holland, D.W. Smith, J. Appl. Phys 61 (12) 5399-5409, 15 June 1987

4. A.I. Ekimov and A.A. Onushchenko, JETP Lett. 34(6), 345-348, (1981)

5. A.I. Ekimov, A.A. Onushchenko, Sov. Phys. Semicond. 16(7), 775-778, July 1982

6. A.I. Ekimov, A.L. Efros, A.A. Onushchenko, Solid State Communications, Vol. 56, No.11, 921-924 (1986)

7. A.I. Ekimov, A.A. Onushchenko, A.G. Plyukhin, A.L. Efros Sov. Phys. JETP 61 (4) 891-897, April 1985

8. D.L. Morse, American Ceramic Society, Vol. 20, No. 3, 777- 780, 1981

9. E. Hanamura, Phys. Rev., B37, 1273, 1988

10. T. Hiroshima, E. Hanamura, Quant. Wells for Optics and Optoelectronics, Technical Digest pp. 92, 1989.

11. Y. Masumoto, M. Yamazaki, H. Sugawara, Appl. Phys. Lett. 53 (16) 1527-1529, 17 October 1988

12. R.J. Ginther, R.D. Kirk, J. Non-Cryst. Solids, Vol. 6, 89 1971

13. S. Parke, R.S. Webb, Physics and Chemistry Glasses, Vol. 13, No. 6, 157-160, December 1972.

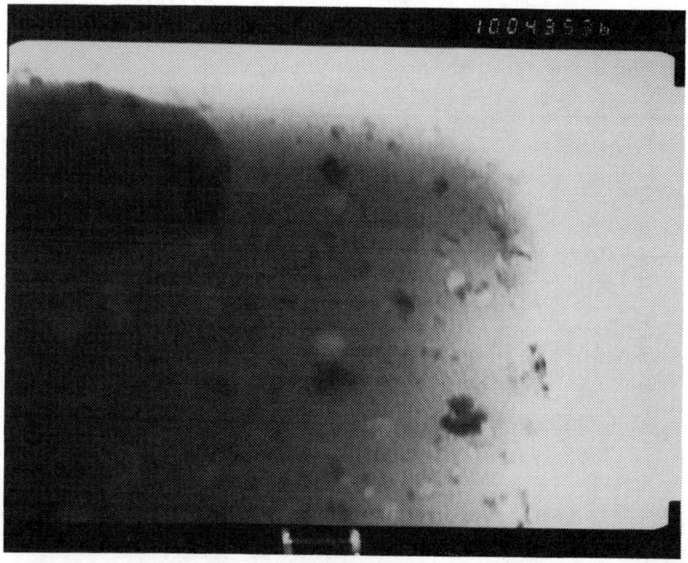

1. TEM photograph of a glass containing CuCl microcrystals that has been heat-treated for 1 hour at 585°C.

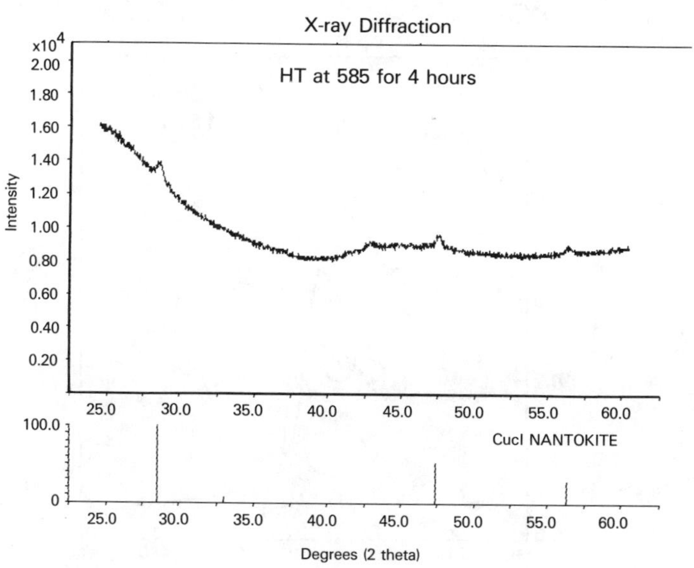

2. X-ray diffraction curve (upper trace) of a heat-treated sample containing CuCl and the peak locations of crystalline CuCl (lower lines).

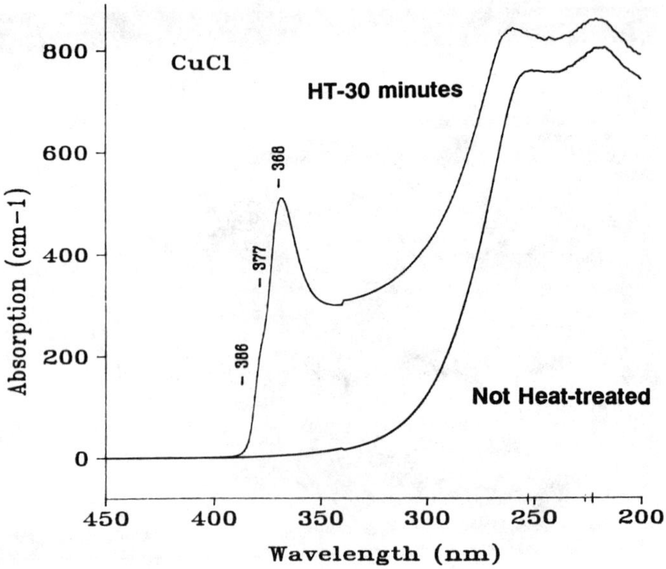

3. Absorption spectra comparing an unheat-treated sample to a glass sample containing CuCl crystals that has been heat-treated for 30 minutes at 585°C.

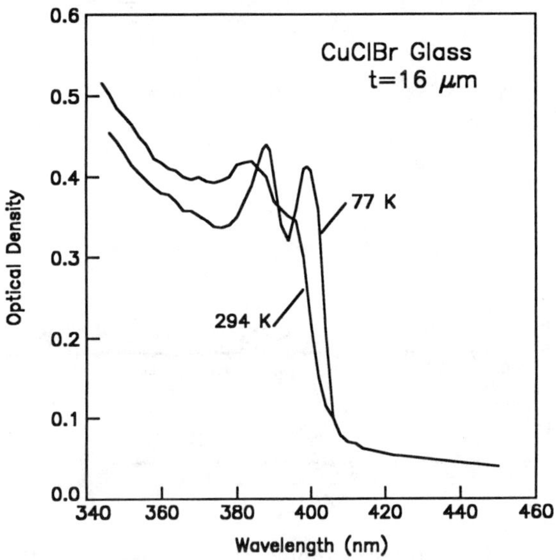

4. Absorption spectra at room temperature and 77K of a 16 μm thick $CuCl_xBr_{1-x}$ sample, heat-treated for 4 hours at 585°C.

Section III. Optical Thin Films and Applications

SOL-GEL PROCESSING OF FERROELECTRIC THIN FILMS

S.L. Swartz, S.J. Bright, and J.R. Busch
Battelle
505 King Avenue
Columbus, OH 43201

T.R. Shrout
Materials Research Laboratory
Pennsylvania State University

ABSTRACT

Ferroelectric thin-film technology is reviewed, and recent results are presented on the sol-gel processing of PZT and PLZT thin films. A general observation has been that the crystallization of sol-gel PZT and PLZT thin films is very sensitive to the substrate material. A sol-gel fabrication process was developed at Battelle for the production of perovskite PZT and PLZT thin films on a wide variety of substrates. This approach involves the deposition of a sol-gel $PbTiO_3$ (or $SrTiO_3$) interlayer between the substrate and the desired PZT or PLZT thin film. The crystallinity of sol-gel PZT and PLZT thin films was improved when the films were deposited on an isostructural perovskite interlayer.

Sol-gel PZT (60/40), PLZT (9/65/35), and composite $PbTiO_3$/PZT and $PbTiO_3$/PLZT thin films were prepared on platinum-coated silicon and [110] sapphire substrates. The crystallization of these PZT and PLZT thin films was influenced by the substrate material and the presence of a $PbTiO_3$ interlayer. PZT and PLZT thin films crystallized with strong [111] orientation when deposited directly on platinum-coated silicon substrates. This orientation was not observed when the $PbTiO_3$ interlayer was present, nor when the PZT and PLZT thin films were deposited on sapphire substrates. Dielectric constants of the PZT, PLZT, and composite $PbTiO_3$/PLZT thin films were low (ranging from 120 to 200), but a relatively high dielectric constant of 1000 was measured on a composite $PbTiO_3$/PZT thin film.

INTRODUCTION

The potential of ferroelectric thin films for electronic and electro-optical device applications has been known for several years [1]. Electronic applications of ferroelectric thin films include nonvolatile memories, thin-film capacitors, pyroelectric sensors, and surface acoustic wave substrates. Electro-optical applications include optical waveguide devices (switches, spatial light modulators), optical memories and displays, and frequency doublers for diode lasers. Important materials for thin-film applications include perovskite ferroelectrics such as $BaTiO_3$, $PbTiO_3$, $Pb(Zr,Ti)O_3$ (PZT), $(Pb,La)(Zr,Ti)O_3$ (PLZT), $KNbO_3$, and $Pb(Mg_{1/3}Nb_{2/3})O_3$, and tungsten-bronze ferroelectrics such as $(Sr,Ba)Nb_2O_6$ and $(Pb,Ba)Nb_2O_6$. Useful ferroelectric thin-film materials and their respective applications are summarized in Table 1.

Much of the development of ferroelectric thin film technology has involved physical deposition methods (e.g., sputtering), and ferroelectric thin-film materials, such as PZT and PLZT [2-6]. However, commercialization of sputtering methods for thin-film ferroelectric applications has been limited by technical difficulties related to achieving high quality and uniform thin films over large areas. These limitations may be reduced as more advanced sputtering methods are developed. Chemical vapor deposition (CVD) and metal-organic chemical vapor deposition (MOCVD) also have been applied to ferroelectric thin films [7-9]. Despite the complicated chemistry associated with MOCVD, this method shows promise for ferroelectric thin film deposition. An additional advantage of MOCVD is that scale-up to large areas and production quantities can be accomplished.

Sol-Gel Processing of Ferroelectric Thin Films

In the past five years, developments in sol-gel processing of ferroelectric thin films have renewed interest in ferroelectric thin-film devices. Commercialization of ferroelectric (PZT) thin films in nonvolatile semiconductor memories is imminent. PZT thin films prepared by sputtering have been utilized in prototype memory devices [10], but sol-gel PZT thin films for nonvolatile memories currently are under development [11-12]. Although, sol-gel ferroelectric thin-film technology currently is driven by these memory applications, optical applications of sol-gel derived ferroelectric thin films will become possible as film quality improves. Numerous institutions currently are involved with

Table 1: Applications of Ferroelectric Thin Films

Thin-Film Materials	Phenomenon	Applications
Perovskite Titanates		
$BaTiO_3$, $(Ba,Sr)TiO_3$	Dielectric Pyroelectric PTCR	Capacitors IR detectors Thermistors
$PbTiO_3$	Pyroelectric Piezoelectric	IR detectors Acoustic transducers
$Pb(Zr,Ti)O_3$ (PZT)	Dielectric Pyroelectric Piezoelectric Electro-optic	Nonvolatile memories IR detectors SAW substrates Waveguide devices
$(Pb,La)(Zr,Ti)O_3$ (PLZT)	Pyroelectric Electro-optic Electro-optic Electro-optic	IR detectors Waveguide devices Optical memories Displays
Perovskite Niobates		
$Pb(Mg_{1/3}Nb_{2/3})O_3$ (PMN)	Dielectric Electro-optic	Capacitors Waveguide devices
$LiNbO_3$	Electro-optic	Waveguide devices
$KNbO_3$	Electro-optic Electro-optic	Waveguide devices Frequency doublers
$K(Ta,Nb)O_3$	Pyroelectric Electro-optic	IR detectors Waveguide devices
Tungsten-Bronze Niobates		
$(Sr,Ba)Nb_2O_6$	Pyroelectric Electro-optic	IR detectors Waveguide devices

research and developmentof sol-gel ferroelectric thin films; a partial listing of these institutions, along with their targeted ferroelectric thin-film materials, is presented in Table 2.

In the sol-gel process, a solution of metal-organic (alkoxide) precursors is prepared with metal cations in the desired stoichiometry. This solution is hydrolyzed, and a thin film is applied to a substrate by spin-coating. The dried film is annealed at relatively low temperatures (500 to 800°C) to remove the organics, and to crystallize and densify the film. Single-layer sol-gel thin films can be deposited with thicknesses of up to 2000 angstroms after annealing; thicker films are produced by multiple deposition/annealing cycles.

Table 2. Organizations involved in Sol-Gel Ferroelectric Thin Film Research

Organization	Materials under Investigation	References
Universities		
Illinois	$PbTiO_3$, PZT, PLZT, PMN	16-17
Penn State	PZT, PLZT, PMN	
Alfred	PLZT	18
New Mexico	PLZT	
Arizona State	PZT	14-15
Tufts	$PbTiO_3$, PZT	19
Queen's (Canada)	PZT, PLZT	20
UCLA	PZT, $KNbO_3$, $(Pb,Ba)Nb_2O_6$	21
Industrial Organizations		
Battelle	$PbTiO_3$, PZT, PLZT, $KNbO_3$	22-23
Raytheon	PZT	
Krysalis	PZT	
Ramtron	PZT	
McDonnell-Douglas	PZT	14-15
Aerospace Corp.	PZT	24
Sandia	PZT, PLZT, $KNbO_3$	25
Westinghouse	$LiNbO_3$	26

Much of the recent sol-gel ferroelectric thin-film research has been directed at the PbO-based perovskite compounds such as $PbTiO_3$, PZT, and PLZT [16-24]. These studies have shown that the crystallinity and microstructure, and thus the ferroelectric properties, of perovskite thin films are highly dependent on the sol-gel processing methods and/or the substrate materials. The film quality is of utmost importance to the optimization of the electrical and optical properties of ferroelectric thin films. Sol-gel thin films must be produced with good crystallinity, and without defects (e.g., pinholes and thickness gradients). In device applications, ferroelectric thin films will be deposited on either conducting (e.g., platinum-coated silicon) or transparent (e.g., sapphire) substrates. Reaction of the substrate material with the ferroelectric film must be prevented and good adherence of the film to the substrate is required.

Sol-Gel Research at Battelle

Battelle has been investigating the sol-gel processing of ferroelectric thin films of $PbTiO_3$, PZT, PLZT, $SrTiO_3$ and $KNbO_3$. The goal of this research is the development of improved materials for electro-optical device applications.

In earlier work at Battelle, it was shown, for a particular set of sol-gel processing and annealing conditions, that the crystallization of sol-gel $PbTiO_3$, PZT and $KNbO_3$ thin films was dependent on the substrate material [22-23]. These results are summarized in Table 3. Of particular interest was the observation of epitaxial crystallization of sol-gel thin films of both PZT and $KNbO_3$ on single-crystal $SrTiO_3$ substrates [22].

Recently, we reported that crystalline PZT and PLZT thin films can be prepared on a wide variety of substrates by first depositing an interlayer (up to 2000 Å) of $PbTiO_3$ (or $SrTiO_3$) between the substrate and the PZT (or PLZT) thin film [23,27]. The crystallization of sol-gel $PbTiO_3$ thin films into the perovskite structure is relatively insensitive to the substrate material. $PbTiO_3$ thus can be used as an interlayer for the more difficult-to-crystallize thin-film materials such as PZT and PLZT. This concept was demonstrated for PZT and PLZT thin films using $PbTiO_3$ and $SrTiO_3$ interlayers and various substrate materials. Perovskite PZT or PLZT thin films crystallized more easily on the isostructural $PbTiO_3$ interlayer than on the original substrate.

With proper control of the sol-gel processing and annealing conditions, crystalline and perovskite-structured PZT and PLZT thin films can be produced without a $PbTiO_3$ interlayer. However,

Table 3. Crystallization of Sol-gel Thin Films on Various Substrates (from Reference 22).

Material	Substrate	Crystalline Phases	Orientation
$PbTiO_3$	fused silica	perovskite	random
	silicon	perovskite	random
	alumina	perovskite	random
PZT (20/80)	fused silica	amorphous	----
	silicon	amorphous	----
	alumina	perovskite	random
	[210] $SrTiO_3$	perovskite	epitaxial
$KNbO_3$	silicon	non-perovskite	----
	alumina	non-perovskite	----
	[210] $SrTiO_3$	perovskite	epitaxial

the use of a $PbTiO_3$ (or other perovskite) interlayer may improve the crystallinity and/or modify the crystallographic orientation of the sol-gel thin films. The use of a perovskite interlayer also may allow for novel composite thin-film geometries to be produced. For example, composite ferroelectric thin films with a gradient index of refraction may possess unique optical properties, whereas gradient-polarization films may be useful for certain electronic applications.

Scope of Present Investigation

Sol-gel $PbTiO_3$, PZT (60/40), and PLZT (9/65/35), and composite $PbTiO_3$/PZT and $PbTiO_3$/PLZT thin films were prepared on platinum-coated silicon and [110] sapphire substrates. The thin films were characterized by x-ray diffraction (XRD) and scanning electron microscopy (SEM). The effects of the $PbTiO_3$ interlayer and the substrate material on the crystallization of sol-gel PZT and PLZT thin films were determined. Dielectric property measurements were made on the sol-gel PZT and PLZT thin films deposited on platinum-coated silicon substrates.

EXPERIMENTAL PROCEDURE

Sol-gel thin films of $PbTiO_3$, PZT (60/40), and PLZT (9/65/35) were produced on platinum-coated silicon and [110] sapphire substrates, by multiple spin-coating and annealing operations. The platinum-coated silicon substrates were prepared by first sputtering 1000 angstroms of titanium on an oxidized silicon substrate, and then sputtering 1000 angstroms of platinum.

A flow diagram describing the preparation of the hydrolyzed alkoxide solutions, the spin-coating deposition, and annealing of $PbTiO_3$, PZT, and PLZT is shown in Figure 1. A complex alkoxide solution was prepared in a methoxyethanol solvent, using a procedure similar to that described by Budd, et al. [16]. The starting chemicals were lead acetate hydrate, titanium isopropoxide, zirconium n-propoxide, and lanthanum isopropoxide. The alkoxide solutions (initially about 0.50 M of perovskite compound) were hydrolyzed with methoxyethanol solutions containing water and nitric acid in amounts corresponding to 2/1 and 0.2/1 moles per mole of perovskite compound. The final alkoxide concentrations corresponded to 0.15 M for the hydrolyzed $PbTiO_3$ solution, and 0.20 M for the PZT and PLZT solutions.

All substrate cleaning, spin-coating, and annealing operations were performed in a class-100 clean room. The substrates were cleaned thoroughly in soapy water, rinsed with distilled water, and blow-dried with nitrogen. This substrate cleaning procedure was found to improve film quality. Hydrolyzed alkoxide solutions were applied to both platinum-coated silicon and [110] sapphire substrates through a filtered (0.2-micron) syringe, and the substrates were spun at speeds of 1000 to 3000 rpm. After each spin-coating deposition, the substrates were placed on a hot plate (at 300°C) for about five minutes. After the final spin-coating deposition (and hot plate treatment), the substrates were annealed in a tube furnace (with flowing oxygen) at a temperature of 600°C for 30 minutes (for the $PbTiO_3$ films) or 700°C for 30 or 60 minutes (for the PZT and PLZT films). Composite $PbTiO_3$/PZT and $PbTiO_3$/PLZT films were produced in the following manner: four layers of $PbTiO_3$ (with intermediate 300°C hot-plate treatments and a final anneal at 600°C for 30 minutes), followed by eight coats of either PZT or PLZT (with intermediate 300°C hot plate treatments and a final anneal at 700°C for 30 minutes). The preparation conditions for $PbTiO_3$, PZT, and PLZT, and composite $PbTiO_3$/PZT and $PbTiO_3$/PLZT thin films are given in Table 4.

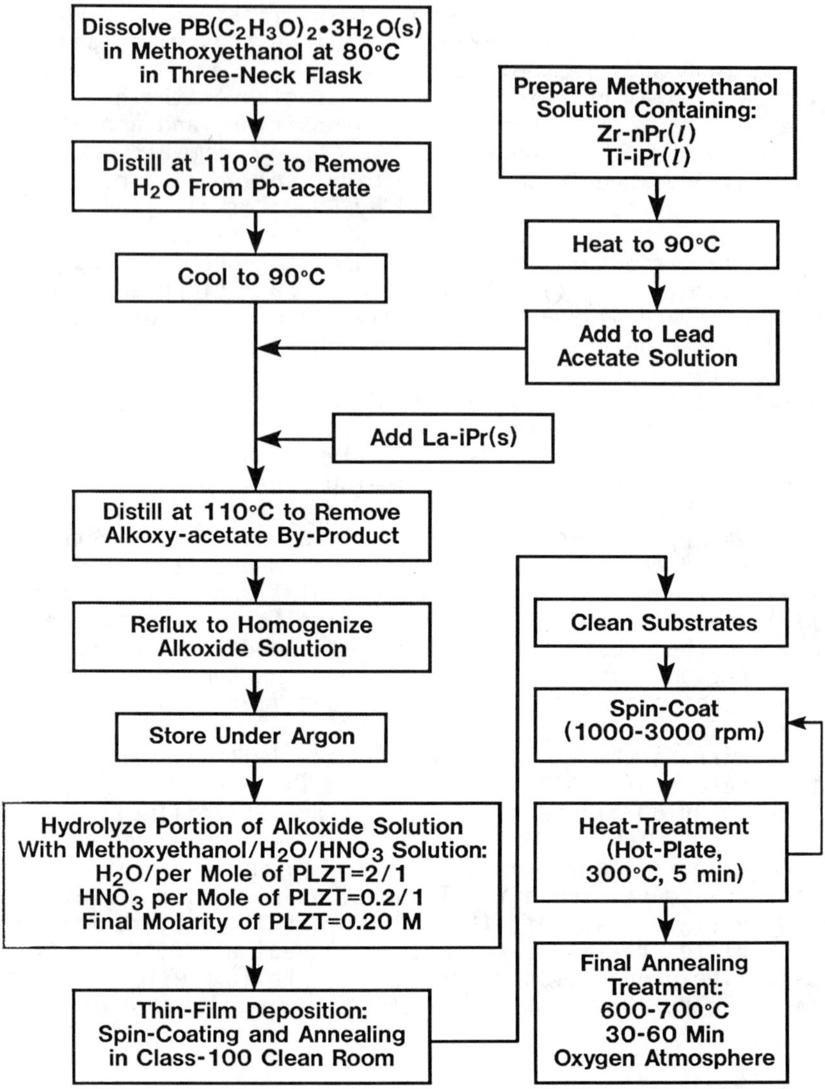

Figure 1. Fabrication Process used for Sol-Gel $PbTiO_3$, PZT, and PLZT Thin Films.

Table 4. Preparation Conditions for Sol-Gel Thin Films

Film	Spin-coating Speed (rpm)	$PbTiO_3$ Coats	PZT Coats	PLZT Coats	Annealing Conditions Temp (°C)	Time (min)
Pt-coated Silicon Substrates						
PT-1	3000	4	0	0	600	30
PZT-1	3000	0	8	0	700	30
PZT-2	1500	0	8	0	700	60
PZT-3	3000	4	8	0	700	30
PLZT-1	3000	0	0	8	700	30
PLZT-2	1500	0	0	8	700	60
PLZT-3	3000	4	0	8	700	30
PLZT-4	1000	4	0	8	700	60
[110] Sapphire Substrates						
PT-2	3000	4	0	0	600	30
PZT-4	3000	0	8	0	700	30
PZT-5	3000	4	8	0	700	30
PLZT-5	3000	0	0	8	700	30
PLZT-6	3000	4	0	8	700	30

X-ray diffraction was used to confirm whether the sol-gel PZT and PLZT thin films had crystallized into the desired perovskite phase, to determine the nature of crystallographic orientation, and to estimate lattice parameters. SEM analysis was performed on fractured cross-sections of the films to obtain grain size and film thickness information. PZT and PLZT thin films deposited on platinum-coated silicon substrates were subjected to dielectric measurements. An array of 1.0-mm diameter gold electrodes was sputtered onto the film surface to act as the top electrode. Contact to the bottom platinum electrode was made with silver paint after scratching through the film. Dielectric measurements were made using an LCR meter (HP-4274A); the AC measurement voltage ranged from 0.01 to 5.0 volts. Electrical resistivity was measured at room temperature with a picoammeter (HP-4192A), by applying 1 volt and measuring the current after 60 seconds.

RESULTS AND DISCUSSION

X-ray Diffraction

XRD patterns of sol-gel thin films of pure PZT (60/40) and composite $PbTiO_3$/PZT thin films on platinum-coated silicon substrates and [110] sapphire substrates are presented in Figures 2a and 2b. XRD patterns of pure PLZT (9/65/35) and composite $PbTiO_3$/PLZT thin films deposited on platinum-coated silicon and [110] sapphire substrates appear in Figures 2c and 2d. The PZT and PLZT thin films both exhibited pseudo-cubic perovskite XRD patterns, whereas a rhombohedral perovskite structure would have been expected. A minor amount of pyrochlore phase (as indicated by a diffraction peak at about 30 degrees), was detected in the sol-gel PLZT thin films deposited on both sapphire and platinum-coated silicon substrates. This pyrochlore phase was absent in the composite $PbTiO_3$/PLZT thin films. Some of the $PbTiO_3$ peaks were observed in the XRD patterns of the composite $PbTiO_3$/PZT and $PbTiO_3$/PLZT thin films, suggesting that solid solution of the $PbTiO_3$ interlayer with the PZT and PLZT thin films did not occur.

The lattice parameters of the tetragonal $PbTiO_3$ interlayer films were calculated from the (100) and (001) diffraction peaks. Lattice parameters of the pseudo-cubic PZT and PLZT thin films were calculated from the (110) diffraction peak. These data are presented in Table 5. The tetragonal lattice parameters of the $PbTiO_3$ interlayer films (after annealing at 600°C) were smaller than those of bulk $PbTiO_3$ (a_o and c_o equal to 3.93 and 4.08 Å for the $PbTiO_3$ interlayer films, compared to a_o and c_o equal to 3.90 and 4.15 Å for single crystal $PbTiO_3$). The lattice parameters of the PZT and PLZT films were identical when deposited on the same substrate, (about 4.080 Å on sapphire and 4.065 Å on platinum-coated silicon substrates). A slight increase in the lattice parameter was observed when the PZT and PLZT films were deposited on the $PbTiO_3$ interlayer. The pseudo-cubic lattice parameters of the PZT and PLZT thin films can be compared with the rhombohedral lattice parameters of about 4.07 and 4.09 Å, for ceramics of PZT (60/40) and PLZT (9/65/35), respectively. The apparent dependency of the perovskite lattice parameter on the substrate material will be important to the optimization of the ferroelectric properties of PZT and PLZT thin films.

Another important observation relates to the crystallographic orientation exhibited by the sol-gel PZT and PLZT thin films. The PZT and PLZT films on platinum-coated silicon substrates exhibited a preferred [111] orientation, apparently resulting from the [111]

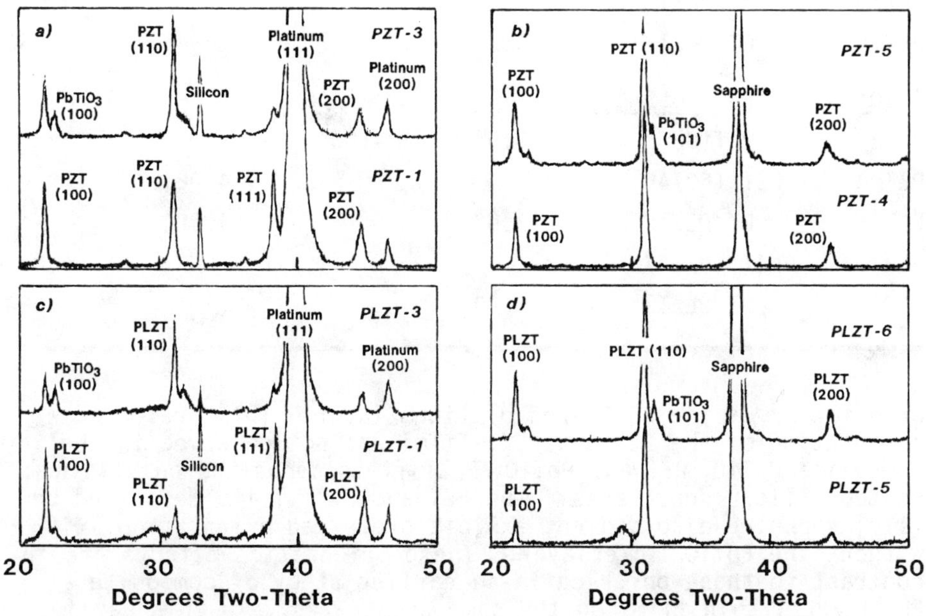

Figure 2: XRD patterns of sol-gel PZT (60/40) thin films (PZT-1 and PZT-4) and composite $PbTiO_3$/PZT thin films (PZT-3 and PZT-5) on (a) platinum-coated silicon and (b) [110] sapphire substrates; and sol-gel PLZT (9/65/35) thin films (PLZT-1 and PLZT-5) and composite $PbTiO_3$/PLZT thin films (PLZT-3 and PLZT-6) on (c) platinum-coated silicon and (d) [110] sapphire substrates.

Table 5. Lattice Parameter Data

Film	Composition	$PbTiO_3$ a_o (Å)	$PbTiO_3$ c_o (Å)	PZT a_o (Å)	PLZT a_o (Å)
Pt-coated Silicon Substrates					
PT-1	$PbTiO_3$	3.92	4.07		
PZT-1	PZT (60/40)			4.066	
PZT-2	PZT (60/40)			4.059	
PZT-3	PT/PZT	3.93		4.066	
PLZT-1	PLZT (9/65/35)				4.059
PLZT-2	PLZT (9/65/35)				4.059
PLZT-3	PT/PLZT	3.94			4.066
PLZT-4	PT/PLZT	3.93			4.066
(110) Sapphire Substrates					
PT-2	$PbTiO_3$	3.94	4.08		
PZT-4	PZT (60/40)			4.081	
PZT-5	PT/PZT	3.93		4.084	
PLZT-5	PLZT (9/65/35)				4.079
PLZT-6	PT/PLZT	3.93			4.084

orientation of the sputtered platinum layer. This preferred orientation of the PZT and PLZT films was not observed in the composite $PbTiO_3$/PZT and $PbTiO_3$/PLZT films deposited on platinum-coated silicon substrates. The PZT and PLZT films deposited on [110] sapphire also did not exhibit preferred orientation, with or without the $PbTiO_3$ interlayer. These orientation effects are in contrast to those observed in an earlier study of composite $PbTiO_3$/PLZT thin films on the same substrates. In this earlier study, strong [100] orientation was observed in composite $PbTiO_3$/PLZT films on sapphire substrates, and only minimal orientation was observed in the same films on platinum-coated silicon substrates [23]. These differences likely are related to the annealing conditions employed. Again, although no definitive explanation can be provided, these orientation effects are of critical importance to the properties of ferroelectric thin films.

Microstructure

SEM micrographs of sol-gel thin films on platinum-coated silicon substrates, PZT-6 (a pure PZT thin film) and PLZT-4 (a composite $PbTiO_3$/PLZT thin film), appear in Figure 3. The PZT-6 film (Figure 3a) was about 0.45 microns in thickness, with a grain size of between 0.1 and 0.15 microns. In the composite $PbTiO_3$/PLZT film (Figure 3b), the $PbTiO_3$ interlayer and the PLZT film could be differentiated, with thicknesses of about 0.25 and 0.45 microns, respectively. The grain size of the $PbTiO_3$ interlayer film was about 0.1 microns, whereas the grain size of the PLZT film was much smaller ($<$0.05 microns). The extremely fine grain size of PLZT compared to PZT was observed throughout this investigation.

Dielectric Properties

Electrical measurements were made with contacts to the bottom (platinum) electrode and individual 1-mm diameter sputtered gold electrodes. Electrical shorts were observed on approximately 20 to 50 percent of the capacitors formed between the bottom platinum electrode and the array of individual gold electrodes. The electrical resistivity (measured at non-shorted regions) was greater than 10^{12} Ω-cm for all films. Measurements of capacitance and dissipation factor (tan δ) were made at several non-shorted points at frequencies between 0.1 and 100 kHz. Dielectric constants of selected films were calculated from the average capacitance and the film thickness (which was estimated from SEM micrographs). These data are presented Table 6, along with data of capacitance and tan δ at frequencies of 0.1, 1, 10, and 100 kHz (measured at a single point). Minimal frequency dispersion of the capacitance and tan δ was observed. Capacitance and tan δ values were relatively constant with AC measurement voltage from 0.01 to 0.5 volts. At higher voltages, the capacitance and loss increased significantly, due to localized heating originating from the extremely high AC field applied to the sub-micron thin films.

Dielectric constants in the range of 1500 to 2000 would be expected for PZT (60/40), and in the range of 3000 to 4000 for PLZT (9/65/35), based on ceramic data. Generally, the dielectric constants observed in the PZT, PLZT, and composite $PbTiO_3$/PLZT

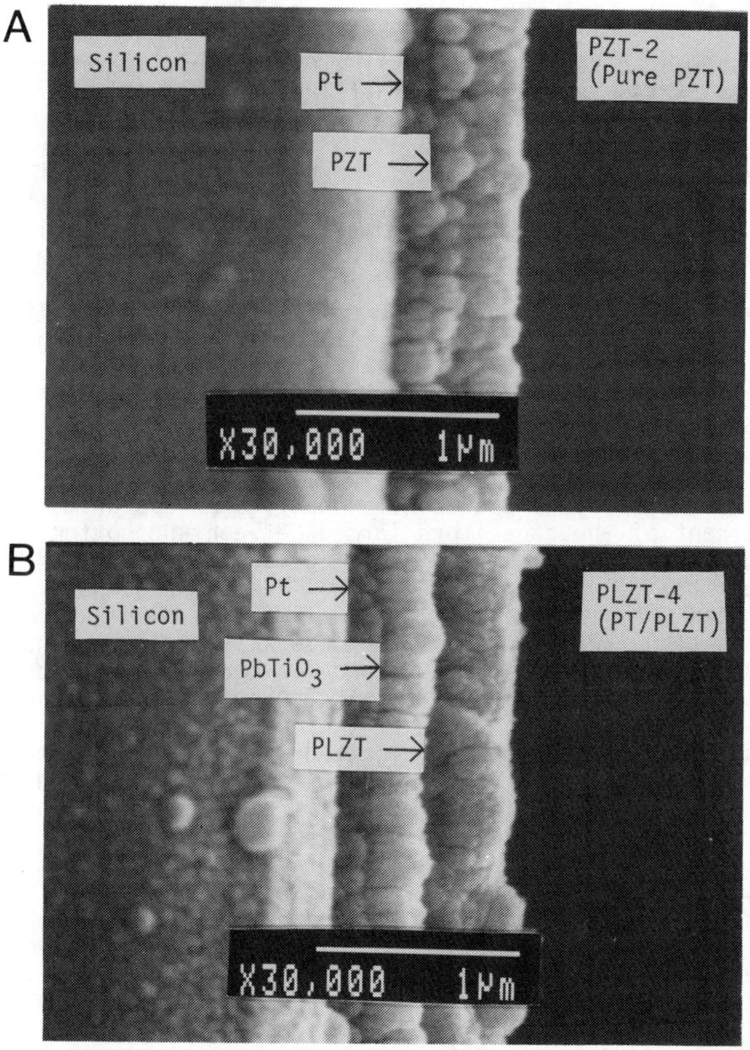

Figure 3. SEM micrographs cross-sections of sol-gel thin films: (a) PZT (60/40) thin film (PZT-2), and (b) composite $PbTiO_3$/PLZT thin film (PLZT-4), deposited on platinum-coated silicon substrates.

Table 6. Dielectric Property Data

Film	Composition	t (μm)	---------Dielectric Data---------			
			K	f (KHz)	C (nF)	tan δ
PZT-2	PZT	0.45	160	0.1	2.52	0.033
				1	2.40	0.034
				10	2.28	0.031
				100	2.19	0.027
PZT-3	PT/PZT	0.40	1000	0.1	19.8	0.080
				1	17.7	0.060
				10	16.4	0.045
				100	15.4	0.060
PLZT-1	PLZT	0.30	200	0.1	5.00	0.056
				1	4.64	0.044
				10	4.40	0.033
				100	4.20	0.028
PLZT-2	PLZT	0.45	120	0.1	1.94	0.045
				1	1.82	0.042
				10	1.72	0.032
				100	1.66	0.025
PLZT-3	PT/PLZT	0.40	200	0.1	3.80	0.047
				1	3.50	0.037
				10	3.40	0.037
				100	3.20	0.038
PLZT-4	PT/PLZT	0.70	190	0.1	1.88	0.035
				1	1.84	0.018
				10	1.78	0.012
				100	1.76	0.012

thin films were much lower, ranging from 120 to 200. However, the dielectric constant measured on the composite $PbTiO_3$/PZT thin film was 1000, significantly higher than that measured on the pure PZT film. The dielectric constant of the composite $PbTiO_3$/PZT film is relatively high, given the low dielectric constant (about 200) of $PbTiO_3$ (which constitutes about one-third of the composite film).

At this time, the exact cause of the low dielectric constants observed in the PZT, PLZT, and PT/PLZT thin films is unknown. Low dielectric constants ($K \ll 1000$) have been observed in PZT thin films prepared by both sputtering [2-4] and sol-gel methods [20]. Dielectric constants in excess of 1000 have been reported for sol-gel PZT thin films [13,17]. The differences may be related to measurement difficulties (electrode problems, interface effects, etc.), or possibly to the combined effects of extremely fine grain size and small film thickness.

SUMMARY AND CONCLUSIONS

A significant amount of current research is devoted to the development of sol-gel thin films of $PbTiO_3$-based perovskites for numerous electronic and electro-optical applications. Important sol-gel thin-film fabrication steps include the preparation of hydrolyzed alkoxide solutions, spin-coating of the solutions onto suitable substrates, and annealing at temperatures in the range of 500 to 800°C. Optimization of the sol-gel processing steps is necessary to achieve the required crystallinity, microstructure, and electrical and optical properties of ferroelectric thin films.

The sol-gel processing of $PbTiO_3$-based perovskite thin films has been under investigation at Battelle. It has been shown that sol-gel $PbTiO_3$ thin films crystallize easily on a wide variety of substrates, whereas the crystallization of PZT and PLZT thin films is relatively substrate dependent. A novel sol-gel fabrication approach was developed for the production of PZT and PLZT thin films. This approach involves the deposition and annealing of an interlayer of $PbTiO_3$ (or $SrTiO_3$), prior to the deposition and annealing of the desired PZT or PLZT thin film. The isostructural perovskite interlayer improves the crystallinity of sol-gel PZT and PLZT thin films, and minimizes the substrate dependency. This perovskite interlayer approach was demonstrated in a study of sol-gel PZT and PLZT thin films, and two-layer $PbTiO_3$/PZT and $PbTiO_3$/PLZT thin films deposited on platinum-coated silicon and [110] sapphire substrates. Key findings of this study are summarized below:

(1) During annealing, sol-gel PZT and PLZT thin films crystallized into the pseudo-cubic perovskite structure. The lattice parameters of both PZT and PLZT thin films were higher (about 4.080 versus 4.065 Å) when deposited on sapphire substrates.

(2) The crystallographic orientation was affected by substrate material and the $PbTiO_3$ interlayer. Strong [111] orientation was observed for sol-gel PZT and PLZT thin films on platinum-coated silicon substrates, whereas relatively little orientation was observed for the PZT and PLZT films on sapphire substrates, or for any of the composite $PbTiO_3$/PZT and $PbTiO_3$/PLZT films on both substrate materials.

(3) The grain size of sol-gel PLZT thin films was smaller than that of the PZT thin film (0.02 to 0.05 versus 0.1 to 0.2 microns).

(4) The resistivities of all films were high ($>10^{12}$ Ω-cm). The dielectric constants were low (between 120 and 200) for the sol-gel PZT, PLZT, and $PbTiO_3$/PLZT thin films, but a relatively high dielectric constant of 1000 was observed for the composite $PbTiO_3$/PZT thin film.

FUTURE RESEARCH DIRECTIONS

This study has demonstrated the importance of sol-gel processing conditions, substrate material, and the existence of a perovskite interlayer to the crystallization of sol-gel PZT and PLZT thin films. Further research, and perhaps new measurement techniques, is required for demonstration and evaluation of the ferroelectric properties of PZT and PLZT thin films. A better understanding of the apparent substrate dependencies and orientation effects observed in sol-gel thin films is required. The relationship between electrode material and device characteristics also must be established. The properties of ferroelectric thin films (e.g., dielectric constant, polarization, ferroelectric hysteresis, piezoelectricity, and aging) must be optimized and tailored to those required for a particular application.

Only preliminary measurements of the electro-optical performance of sol-gel ferroelectric thin films have been reported in the literature. The electro-optic coefficients of sol-gel PZT

(50/50) thin films have been calculated from longitudinal measurements, i.e., through the plane of the film [28]. Optical waveguiding through the plane of a sol-gel PLZT thin film recently was demonstrated [17]. The optimization of the electro-optical properties of ferroelectric thin films will require improvement of the film quality (e.g., reduced defect density) and will require that all fabrication steps be carried out under ultra-clean conditions. Compositional adjustment may be required to further optimize the electro-optical properties of ferroelectric thin films. For instance, it is likely that the electro-optical performance (and performance requirements) of PLZT in thin-film form will be different than that of transparent PLZT ceramics of the same composition. Further, transparency is much easier to achieve in thin films than in ceramics. Certain ferroelectric materials, that are difficult to produce as transparent ceramics (e.g., $PbTiO_3$, $(Pb,La)TiO_3$, and PZT), thus can be considered for electro-optical thin-film applications.

REFERENCES

1. M.H. Francombe, "Ferroelectric Thin Films and their Device Applications", Thin Solid Films **13**, 413-433 (1972).

2. S.B. Krupanidhi, et al., RF Planar Magnetron Sputtering and Characterization of Ferroelectric $Pb(Zr,Ti)O_3$ Films", J. Appl. Phys. **54** (11), 6601-6609 (1983).

3. K. Sreenivas and M. Sayer, "Characterization of $Pb(Zr,Ti)O_3$ Thin Films Deposited from Multi-Element Metal Targets", J. Appl. Phys. **64** (3), 1484-1493 (1988).

4. R. Takayama and Y. Tomita, "Preparation of Epitaxial $Pb(Zr_xTi_{1-x})O_3$ Thin Films and their Crystallographic, Pyroelectric, and Ferroelectric Properties", J. Appl. Phys. **65** (4), 1666-1670 (1989).

5. M. Isheda, et al., "Preparation and Properties of Ferroelectric PLZT Thin Films by RF Sputtering", J. Appl. Phys. **48** (3), 951-953 (1977).

6. H. Adachi, et al., "Ferroelectric $(Pb,La)(Zr,Ti)O_3$ Epitaxial Thin Films on Sapphire Grown by RF-Planar Magnetron Sputtering", J. Appl. Phys. **60** (2), 736-741 (1986).

7. M. Kojima, et al., "Chemical Vapor Deposition of $PbTiO_3$ Thin Film", Jpn. J. Appl. Phys. **22** (22-2), 14-17 (1983).

8. B.S. Kwak, et al., "Metalorganic Chemical Vapor Deposition of $PbTiO_3$ Thin Films", Appl. Phys. Lett. **53** (18), 1702-1704 (1988).

9. S.L. Swartz, et al., "Characterization of MOCVD $PbTiO_3$ Thin Films", Ferroelectrics **93**, 37-43 (1989).

10. D. Bondurant, "Ferroelectronic RAM Memory Family for Critical Data Storage", Proceedings of the First Symposium on Integrated Ferroelectrics, CMC-89 (Colorado Springs, March 31, 1989), 212-215 (1989).

11. J.F. Scott, et al., "Switching Kinetics of Lead Zirconate Titanate Submicron Thin-Film Memories", J. Appl. Phys. **64** (2), 787-792 (1988).

12. J.F. Scott, et al., "Radiation Effects on Ferroelectric Thin-Film Memories: Retention Failure Mechanisms", J. Appl. Phys. **66** (3), 1444-1453 (1989).

13. S.K. Dey and R. Zuleeg, "Processing and Parameters of Sol-Gel PZT Thin Films for GaAs Memory Applications", Proceedings of the First Symposium on Integrated Ferroelectrics, CMC-89 (Colorado Springs, March 31, 1989), 189-194 (1989).

14. S.K. Dey and R. Zuleeg, "Processing and Properties of Sol-Gel PZT Thin Films on GaAs", Proceedings of the 7th Int. Meeting on Ferroelectricity, to appear in Ferroelectrics.

15. J. Harris, Krysalis Corporation, private communication (October 1989).

16. K.D. Budd, et al., "Sol-Gel Processing of $PbTiO_3$, $PbZrO_3$, PZT, and PLZT Thin Films", Br. Ceram. Proc. **36**, 107-121 (1985).

17. K.D. Budd and D.A. Payne, "Sol-Gel Processing of Thin-Layer Dielectrics in Lead Titanate-Based Systems", to be published in <u>Materials for Non-Linear and Electro-optics</u> (Institute of Physics, UK, 1989).

18. V.K. Seth and W.A. Schulze, "Fabrication and Characterization of Ferroelectric PLZT 7/65/35 Ceramic Thin Films and Fibers", Proceedings of the First Symposium on Integrated Ferroelectrics, CMC-89 (Colorado Springs, March 31, 1989), 175-184 (1989).

19. C. Chen and D.F. Ryder, "Synthesis and Microstructure of Highly Oriented Lead Titanate Thin Films Prepared by a Sol-Gel Method", J. Am. Ceram. Soc. **72** (8), 495-498 (1989).

20. Y. Guanghua, et al., "Preparation of $Pb(Zr,Ti)O_3$ Thin Films by Sol Gel Processing: Electrical, Optical, and Electro-optic Properties", J. Appl. Phys. **64** (5), 2717-2724 (1988).

21. X. Yuhuan, et al., "Self Biased Heterojunction Effect on Ferroelectric Thin Film on Silicon", Proceedings of the 7th Int. Meeting on Ferroelectricity, to appear in Ferroelectrics (1989).

22. S.L. Swartz, et al., "Ferroelectric Thin Films by Sol-Gel Processing", Mat. Res. Soc. Proc., Vol. 152, 227-232 (1989).

23. S.L. Swartz, et al., "Sol-Gel Processing of Composite $PbTiO_3$/PLZT Thin Films", Proceedings of the 7th Int. Meeting on Ferroelectricity, to appear in Ferroelectrics (1989).

24. R.A. Lipeles, et al., "Sol-Gel Processing of Lead Zirconate Titanate Thin Films" in Ultrastructure Processing of Ceramics, Glasses, and Thin Films, 320-326 (1986).

25. B.A. Tuttle, et al., "Sol-Gel Processing of $KNbO_3$ Thin Films", Mat. Res. Soc. Proc., Vol. 152 (1989).

26. D.P. Bartlow and J. Greggi, "Properties and Microstructure of Thin $LiNbO_3$ Films Prepared by a Sol-Gel Process", J. Mater. Res. **2** (5), 595-605 (1987).

27. S.L. Swartz and P.J. Melling, "Improved Perovskite Thin Films", U.S. Patent Application (1989).

28. C.E. Land, "Longitudinal Electrooptic Effects and Photosensitivities of Lead Zirconate Titanate Thin Films", J. Am. Ceram. Soc. **72** (11), 2059-2064 (1989).

DEPOSITION OF ELECTROOPTIC THIN FILMS †

Angus I. Kingon*, Shang-Hsien Rou, Michael S. Ameen,
Thomas M. Graettinger, Kenneth Gifford and Orlando Auciello+
Department of Materials Science and Engineering
North Carolina State University, Raleigh, NC 27695-7907

and

Alan R. Krauss
Chemistry and Materials Science Divisions, Argonne National
Laboratory, Argonne, Illinois 60439

ABSTRACT

$KNbO_3$ thin films have been deposited on Si and MgO (100) substrates by a new ion beam sputter deposition system. It has been shown that, by choice of substrate and conditions, a variety of microstructures can be controllably achieved, varying from polycrystalline to epitaxial single crystal. In the case of the latter, two defect types have been observed, viz antiphase domains and 211-type growth twins. A structure has been postulated for the domain boundaries of the former. The relationships between microstructures and ferroelectric properties can now be investigated. Preliminary dielectric property measurements were made of polycrystalline films deposited on TiN / MgO (100) substrates. In the case of epitaxial $KNbO_3$ films on MgO substrates, light has successfully been coupled into the films, and the films show low scattering losses.

* Member, American Ceramic Society
+ Also Microelectronics Center of North Carolina
† Invited Presentation

1. INTRODUCTION

1.1 Background to Electrooptic Thin Films

With the present trend towards developing smaller, integrated optical devices, **thin film** electrooptic components, such as guided wave phase modulators for optical switches, appear increasingly attractive. Electrooptic components fabricated by thin film process methods display several immediate and potential advantages, which include: processing techniques amenable to a wide range of device architectures; potentially fewer process steps and lower costs; scale of film thicknesses compatible with requirements for integrated optic devices; lower growth temperatures; (economic) tolerance for lower growth rates (thereby a potential for increased material quality); and the possibility of "graded" compositions and tailoring of refractive index profiles.

1.2 $KNbO_3$ as an Electrooptic Material

The material commonly utilized for phase modulators is $LiNbO_3$, although it does not have the highest figure of merit.[1,2] Rather, its utilization is due to its convenient fabrication in bulk single crystal form by growth from the melt. In addition, channel waveguides of cross-sectional dimensions which match optical fibers are conveniently formed by Ti ion-diffusion into the $LiNbO_3$ wafers.

Materials with higher figures of merit exist, although a precise comparison necessitates definition of the device geometry. These materials include[1,2] $KNbO_3$ (33- and 42- modes), $BaTiO_3$ (51- and 33- modes, and $Ba_2NaNb_5O_{15}$ (33- mode).

$KNbO_3$ melts incongruently, making growth of bulk, high quality single crystals difficult.[3] It has a Curie point of ~435°C, and undergoes a further tetragonal-orthorhombic transition at ~225°C.[4] The structural anisotropy promotes micro- and macrocracking during cooling through the phase transitions.

It is interesting to note that in the K_2O - Nb_2O_5 system, a number of compounds have been reported, ranging in stoichiometry from $K_6Nb_{44}O_{113}$ to $K_6Nb_2O_8$.[5] In addition, deviation from the ideal cation stoichiometry must be expected. The extent of this non-stoichiometry, of the form $K_{1-2x}NbO_{3-y}$, has not been investigated.

$KNbO_3$ is therefore a candidate thin film electrooptic material, which requires further investigation. This investigation must include the selection of suitable process methods.

1.3 Thin Film Process Methods

A number of processing methods are currently being explored for ferroelectric thin films, as discussed in an earlier paper.[6] Critical issues determining the choice include: the processing temperature limits; substrates and interfaces; defects and microstructures; and variability of composition and properties. These are briefly mentioned in turn.

Processing temperatures are determined largely by the devices required. In the case where electrooptic components need to be directly deposited on Si and GaAs semiconductors, the process temperatures are limited to typically <600°C. At these reduced temperatures the mobility of cation species is low, making the fabrication of high quality single crystal thin films difficult.

The choice of substrate is determined by the device architecture, the required refractive index mismatch, and the lattice match between substrate and film (if epitaxial films are required). Direct deposition on Si and GaAs presents the problem of the lattice mismatch with $KNbO_3$, and the formation of amorphous oxide layers at the interface.

For most electrooptic applications, it is likely that epitaxial single crystal films will be required, as grain boundaries in polycrystalline films can act as scattering centers, and as a source of defects. This postulate has yet to be confirmed. Lattice defects can yield optical centers and photocarriers, with the resultant transient currents being detrimental to the electrooptic device performance. Unfortunately, detailed relationships between lattice defects and properties, and more generally between microstructures and properties, have not been established for these materials. This must remain a topic of research.

The variability of properties from sample-to-sample and batch-to-batch is an issue central to commercialization of these materials. Tight constraints are imposed on processing, with the control of cation stoichiometry a well-identified problem.

The two techniques most commonly utilized for the deposition of electrooptic and other ferroelectric films are sol-gel processing (or a variant thereof), and magnetron sputtering. While both techniques possess a set of valuable advantages, it is important to be aware of their respective disadvantages.

Sol-gel films usually utilize spin-on deposition of organometallic precursors, with film thicknesses ≥ 0.3 μm requiring multiple depositions. Problems normally encountered are film cracking and porosity, which can be attributed to large volume changes upon drying, or upon pyrolysis and loss of the organic component. The process makes nucleation and growth of the crystalline films difficult to control in comparison with a vapor deposition process, and thus there is difficulty in producing epitaxial films at comparably low temperatures.

Magnetron (or plasma) sputtering has as a major disadvantage in the fact that the plasma interacts directly with the growing film. In the case of oxide deposition, this results in the so-called "negative ion effect"[7], as well as electron bombardment [8], and inert gas incorporation. Methods which minimize the negative ion effect [9], inevitably have a large negative influence on the film growth rate.

In addition, it is common in plasma sputtering to use single multicomponent oxide targets. The problem of stoichiometry control in the presence of preferential sputtering effects has been documented previously.[10] While these effects can be minimized in the short term, the prognosis for commercial-scale processes must be carefully weighed.

Because of the disadvantages mentioned above, new processing methods for multicomponent oxides are presently being explored. In this paper we describe the application of a computer-controlled ion beam sputter deposition technique to $KNbO_3$ thin films, and demonstrate the extent to which microstructures may be controlled.

2. EXPERIMENTAL

2.1 Computer Controlled Ion Beam Sputter Deposition

The system is shown in Figure 1. The deposition system consists of a Nb-lined stainless steel chamber turbopumped to a base pressure of about 1.0 x 10^{-7} torr, an Ar^+ ion beam of 1.4 keV and 25 mA, and a quartz lamp heater assembly capable of heating the substrate to > 600°C. The system geometry shown in Figure 1 was chosen based on previous studies performed on Ar^+ backscattering and incorporation into

COMPUTER - CONTROLLED ION BEAM
DEPOSITION SYSTEM

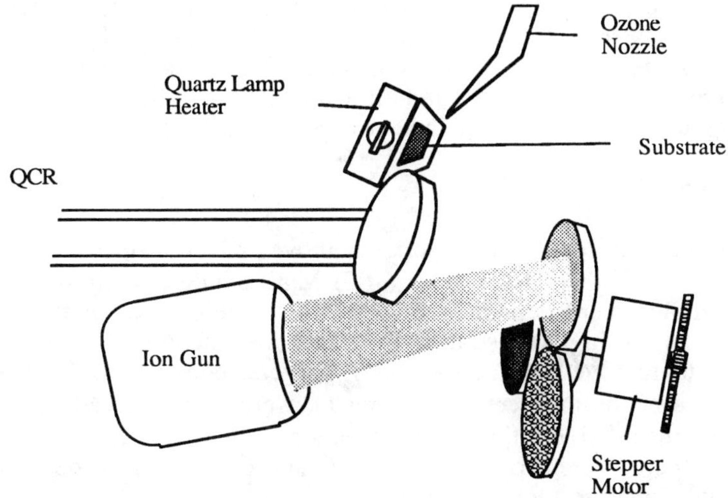

Fig. 1. Schematic of the computer controlled ion beam sputter deposition system.

the growing film.[11,12] These studies indicate that, for Ar^+ ions, near normal beam incidence on the target with the substrate positioned close to the beam axis results in a minimization of ion scattering from the target and thus of gas incorporation into the growing film. Further reduction of deleterious ion scattering effects can be achieved by using source ions of higher atomic mass, such as Kr^+ and Xe^+.[11,12] The quartz crystal resonator (QCR) and substrate are positioned symmetrically about the beam-target (mirror) plane to ensure accurate measurement of the sputtered flux.

A typical deposition cycle proceeds in the following manner: After sputtering from a target, the computer turns the ion beam off, then activates the stepper motor, which rotates a new target into position in front of the ion beam. Switching the beam off in this manner minimizes possible contamination of the film that may arise from sputtering of the holder material during the transition from one target to another. As a precaution the holder is constructed of niobium. The QCR is then automatically programmed by the computer to the density value of the next target material, the thickness reading is reset to zero, and the ion beam is switched on to sputter the target. The thickness of the layer is monitored until the programmed set point is reached. This is presently done with less than 1% variation in layer thicknesses. Typical layer thicknesses are 1 - 20Å at the substrate. The cycle is repeated with each component until the desired film thickness is obtained. [Further details related to this novel method will be published elsewhere.][13]

2.2 Deposition of $KNbO_3$

Using the technique described above, $KNbO_3$ films have been deposited onto either (100) Si or (100) MgO single crystal substrates. Some samples were deposited on TiN films which had in turn been deposited on the MgO substrates. The MgO was chemically cleaned and annealed to remove surface damage. Targets, constructed from KO_2 pressed in a N_2 atmosphere and Nb metal, were fitted into Nb retaining rings and installed on the rotating platen shown in Figure 1. The substrates were heated to between 450°C and 600°C during deposition, and cooled slowly to room temperature.

For the films reported in this study, computer set points for the individual layer thicknesses were 10Å for the Nb and 20Å for each of the two KO_2 targets. These layers interdiffuse at the elevated substrate temperatures to yield $KNbO_3$ in situ. Investigations on reduced layer thicknesses are currently underway. Depositions consisted of approximately 75 rotation cycles, yielding films about 1500Å in thickness. It is assumed that the deposited potassium species is K_2O. Films were transparent with a very slight yellow coloration (typical of $KNbO_3$ single crystals).

2.3 Film Characterization

Films were characterized by x-ray diffraction using Cu $K\alpha$ radiation, and by cross-sectional and plan view transmission electron microscopy using a Hitachi H800. High resolution electron microscopy was performed on a JEOL 200CX. Special TEM specimen preparation procedures have been developed, which will be reported elsewhere.[14]

For samples deposited on TiN / Mgo substrates, impedance was measured over the frequency range 5-1000 kHz using a Hewlett-Packard 4192A impedance analyzer.

$KNbO_3$ was deposited on Si (100), MgO (100) and TiN / MgO (100) substrates at substrate temperatures in the range 450 - 600°C.

3. RESULTS AND DISCUSSION

Over a small range of K:Nb ratios, the desired single phase perovskite $KNbO_3$ resulted. Excess K_2O appeared to be evaporated, with the surface morphology showing some dependence upon this K_2O excess.

Despite varying the K:Nb ratio over a wide range, the only other phase detected was KNb_3O_8. This is in marked contrast to the reported phase diagram which was determined using K_2CO_3, in the presence of atmospheric water vapor.[5] Noting the absence of atmosphere in the present vacuum system and the short interdiffusion distances required for reaction, it is deduced that a number of the previously reported phases are stabilized by H_2O and / or CO_2.

A major objective of the present work was to determine the range of microstructures which could be achieved. Table 1 summarizes the results for Si and MgO substrates.

Table 1 Summary of $KNbO_3$ Microstructures Observed

Substrate	Substrate Temperature (°C)	Microstructures	Figures numbers
Si	500	amorphous	
	550	microcrystalline, amorph. matrix	
	600	polycryst., little orientation	2
MgO	500	amorphous	
	550	mixture non-oriented and epitaxial	3
	600	epitaxial,mostly single crystal	4,6,7
TiN/MgO	600	polycrystalline	

Selected microstructures are shown in Figures 2-7. Figure 2 is a cross-sectional TEM micrograph of $KNbO_3$ deposited on Si (100) at 600°C. It is apparent that there is a significant formation of amorphous SiO_2 at the interface. EDS analysis showed no measurable concentration of potassium in this interface layer. The polycrystallinity is an expected result of the amorphous interface. Analysis of further samples indicates that the thickness of the amorphous SiO_2 layer is variable, and dependent upon process conditions. The surface roughness of the $KNbO_3$ films on silicon is a function of the grain size, which is is ~ 0.5 - 1.0 μm at 600°C deposition temperature.

Figure 3 shows a plan view TEM micrograph of $KNbO_3$ deposited on MgO (100) at 550°C. The film is polycrystalline, with some grains displaying an orientational relationship (dark grains) to the substrate, and with others showing no apparent correlation. This result is confirmed by x-ray diffraction. It is interesting to note that at the same deposition temperature, films on Si are predominantly amorphous, with some randomly aligned microcrystalline regions, about 20 - 30 nm in diameter. This emphasizes the role of the substrate on the nucleation and growth process.

Figures 4, 6 and 7 show microstructures of $KNbO_3$ deposited on MgO (100) single crystal substrates at 600°C. X-ray diffraction and selected area electron diffraction indicate that the whole film has an epitaxial relationship to the substrate.

The film appears to have tetragonal symmetry, rather than orthorhombic. Selected area diffraction (SAD) of cross-sectional TEM samples indicates that the c-axis is oriented perpendicular to the plane of the film. It should be emphasized that the film is epitaxial to the substrate, and not simply oriented, as the a and b axes of the film match directly onto the a and b axes of the cubic MgO substrate. Figure 4 shows a high resolution lattice image of a film in plan view. The measured interplanar spacings confirm the lattice parameters of a (=b) = 4.0 ± 0.05 Å . High resolution images also indicate that there are two types of defect types present in these single crystal films, viz antiphase domain boundaries (APDs) and growth twins. The antiphase domain boundaries represent stacking faults in the cation sublattice, i.e., a transposition of A and B sites across the boundary, and a shift of one half of a unit cell. An APD has been outlined in Fig. 4. The boundary encloses a volume element. The concentration of these APDs varies

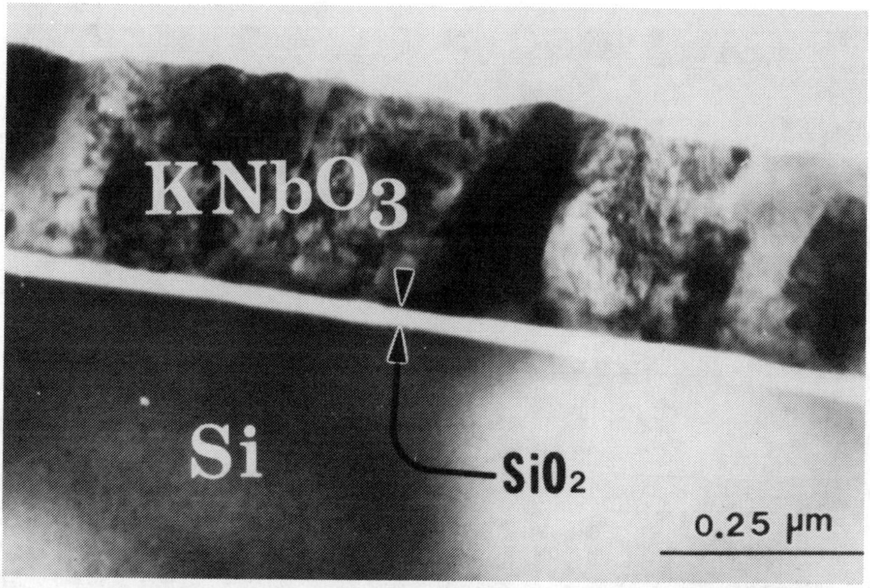

Fig. 2. Cross-sectional TEM image of $KNbO_3$ deposited on Si (100) at 600°C.

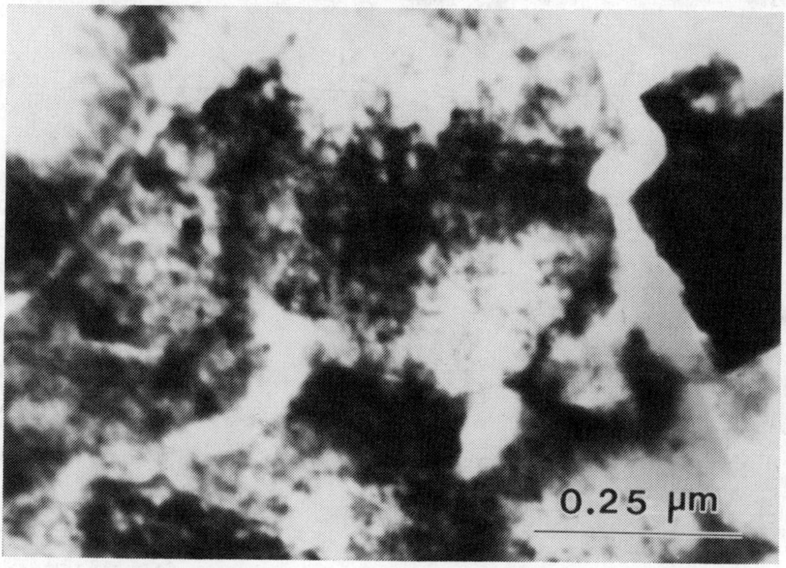

Fig. 3. Plan view TEM image of $KNbO_3$ deposited on MgO (100) at 550°C. SAD indicates the dark grains to be epitaxial to the substrate.

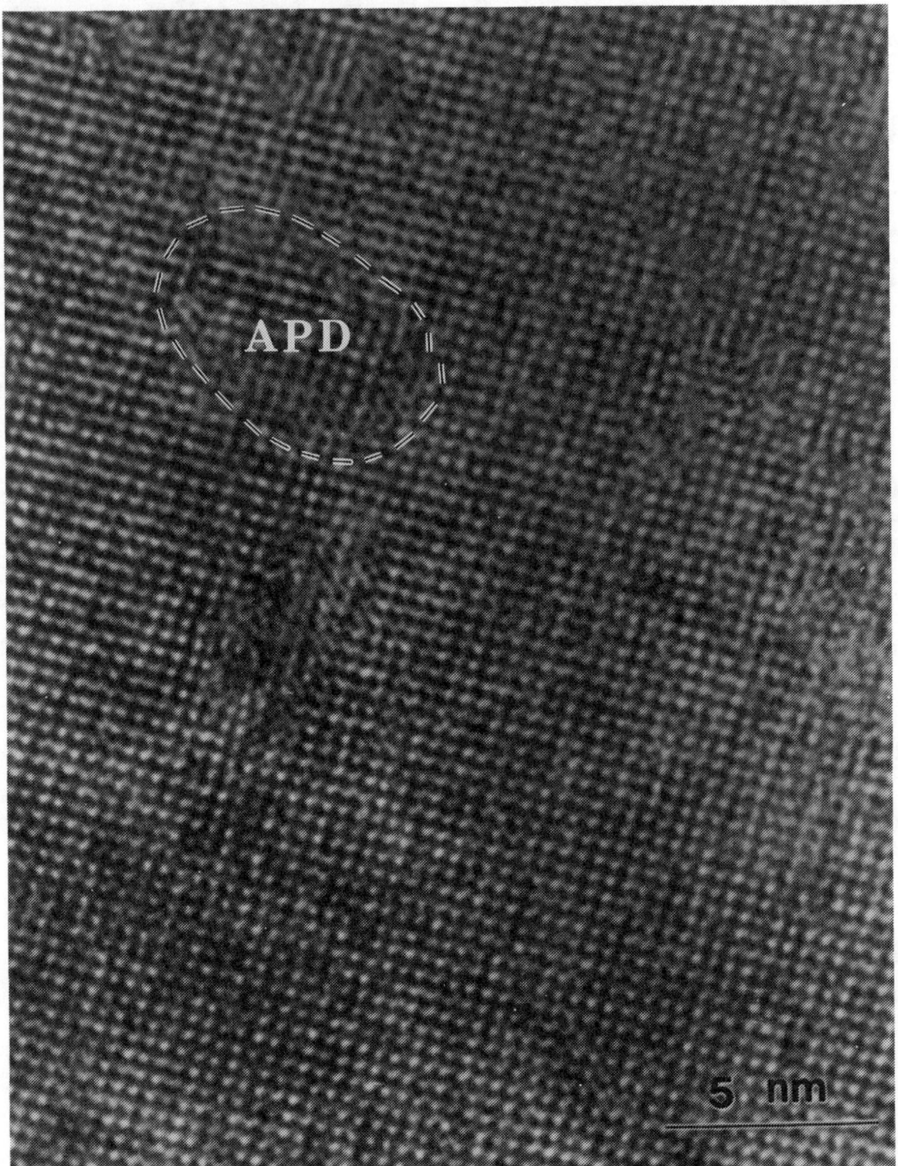

Fig. 4. High resolution electron microscope image of epitaxial $KNbO_3$ on a MgO (100) substrate. An APD is outlined.

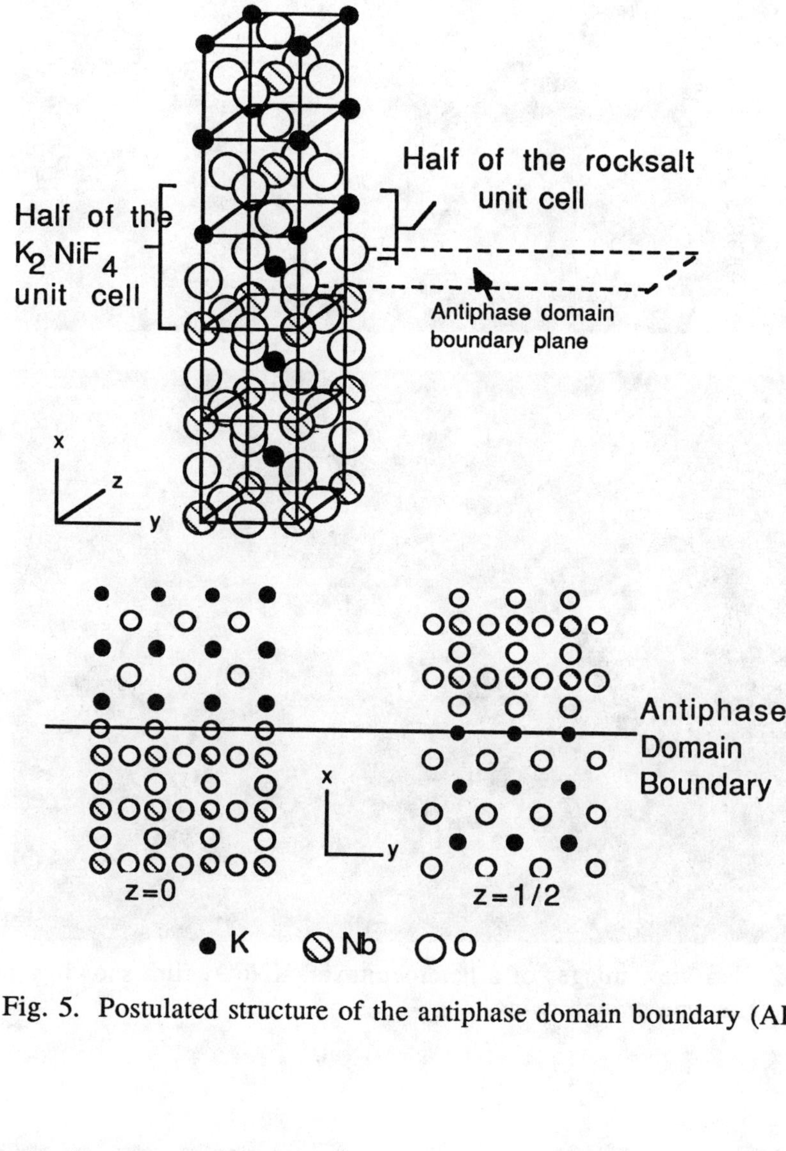

Fig. 5. Postulated structure of the antiphase domain boundary (APD).

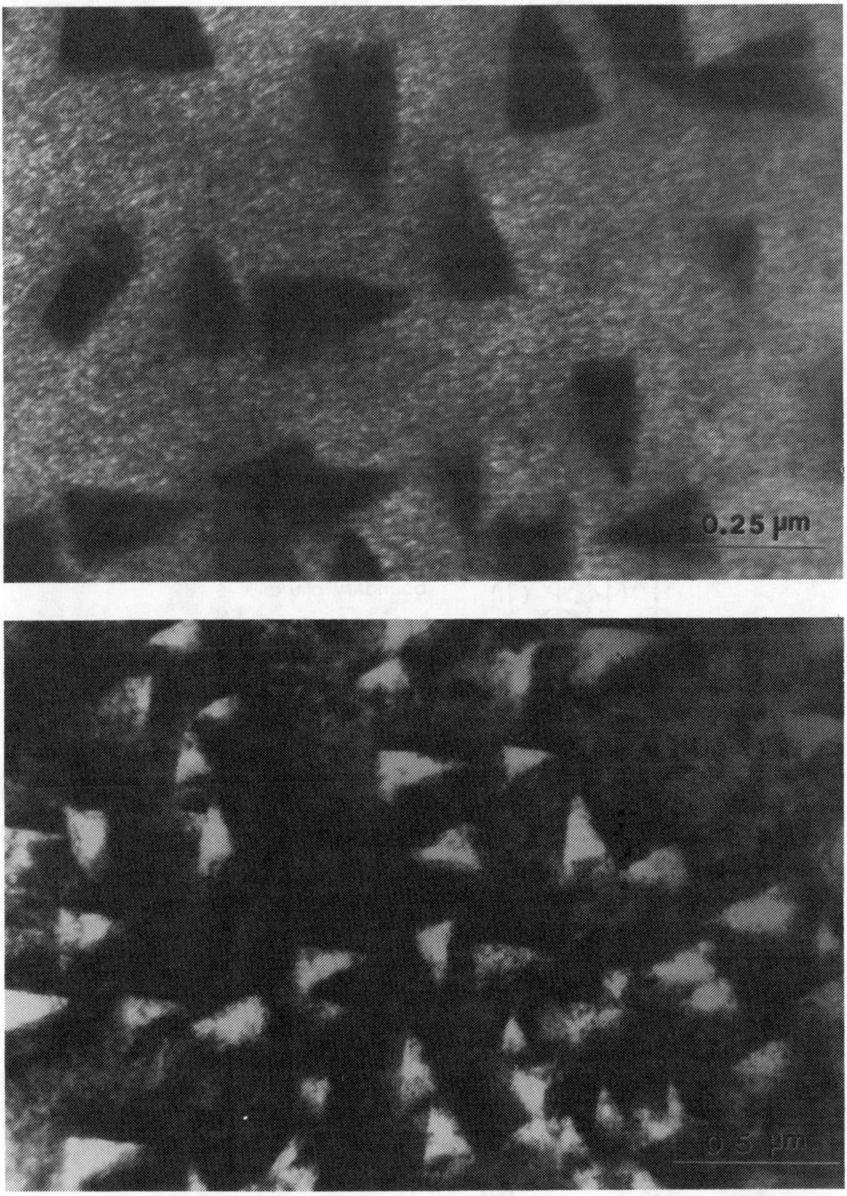

Fig. 6. Plan view images of a heteroepitaxial $KNbO_3$ film showing the 211-type growth twins.

both within one film, and from film-to-film. The origin of the stacking faults could either be due to nucleation of growth at various points within the film, or (if nucleation occurs at the interface as is desired), then due to steps and ledges on the MgO substrate surface.

A postulated structure for the boundary is shown in Figure 5. This postulated structure is expected to be of low free energy. The niobium retains coordination to six oxygens in an octahedral site, while the structural inversion occurs around potassium, with four of its adjacent oxygens in "next nearest neighbor" sites being replaced by a single oxygen in a "nearest neighbor site". This is considered acceptable for the more ionic bonds to the A-site K^+ cation, as a local rocksalt-type structure results right at the boundary. The postulate that this is a low energy boundary is strengthened by the fact that the structure emulates half the unit cell of the K_2NiF_4 structure, which is a common variant of the perovskite structure.[15] However, it should be noted that this postulated structure is less favourable for $A^{1+}B^{5+}O_3$ perovskites. This is due to the fact that in the case of the former, there is a net charge associated with each plane of the {200} type, alternating in sign, for a net zero charge. In the case of $A^{2+}B^{4+}O_3$ perovskites, each plane is a zero net charge. Therefore, introduction of a APD as shown in Fig. 5 results in a disruption in the alternating charges in $KNbO_3$, making it significantly less favourable than for the case of $A^{2+}B^{4+}O_3$ structures. The method of compensating for the excess charge at the APD boundary is $KNbO_3$ is not known, but is presumably by oxygen vacancies. Computer simulations of the possible structures, including that shown in Fig. 5, are presently being undertaken.

The other type of defect observed is a growth twin. This is being reported in more detail elsewhere.[16] Figures 6a and 6b are plan view images of these twins. These growth twins have twin boundaries of the {211} type. The twins are therefore of inverted tetrahedral shape, with the triangular base having angles of 71, 71 and 38°. The twins can be pointed in one of four orthogonal directions. A portion of a twin, the "point", is shown in a plan view high resolution atomic image in Fig. 7. The boundary is curved rather than being "sharp" down to atomic dimensions in order to maintain charge stoichiometry.

In the case of these growth twins, it should be noted that the allowed directions of the polarization vectors differ from those of the matrix. This will affect the overall dielectric displacement. The effect on domain wall mobility and thus on switching processes, has not been established.

Initial characterization of the dielectric properties of the films deposited on TiN / MgO has been undertaken. Measurements were made between silver electrodes sputter deposited on the top surface, and the TiN electrode at the $KNbO_3$ lower surface. These particular films were polycrystalline. (Details will be reported elsewhere.[17]) No dielectric breakdown was observed at the maximum applied fields of 1×10^8 Vm^{-1} (1000 kV per cm). The relative dielectric constant decreased from ~100 to ~55 over the frequency range 1 kHz to 1 MHz. However, films were lossy, with tan σ about 0.1. Although the TiN electrode contained some oxygen, no insulating interface (for example $MgTiO_4$ or TiO_2) was observed at the TiN - $KNbO_3$ interface.[17]

In the case of the epitaxial films deposited directly onto MgO substrates, the coupling of light directly into the film was achieved. The configuration is shown schematically in Figure 8. For thicker films (~0.5 μm), up to four coupling modes were observed. Relatively little scattering occurs in the film. The analysis of the coupling modes is presently being undertaken, and will be extended to direct measurement of in-plane transmission losses and electrooptic coefficients.

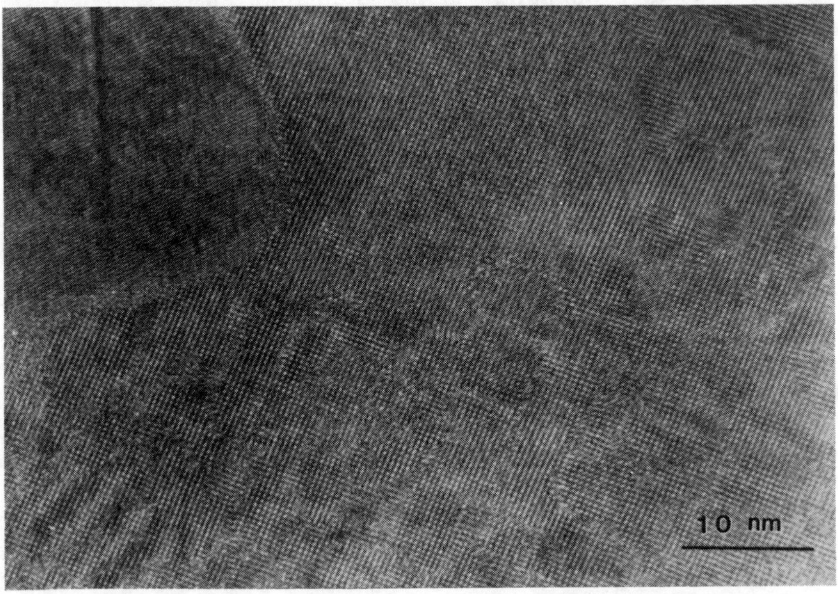

Fig. 7. High resolution image of epitaxial $KNbO_3$ film in plan view, showing "point" of growth twin of the type seen in Figs. 6a and 6b.

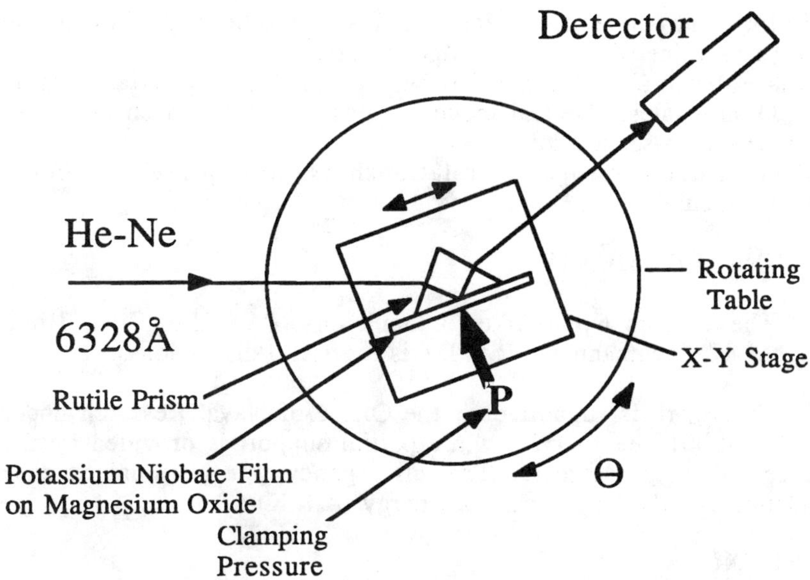

Fig. 8. Schematic of the method for coupling light from a laser source into the plane of the film.

4. CONCLUSIONS

The conclusions of this work may be summarized as:

a) A new computer-controlled ion beam sputter deposition technique has been applied successfully to the synthesis of $KNbO_3$ thin films.
b) Phases observed are $KNbO_3$ and KNb_3O_8, significantly fewer than previously reported in a study of the K_2CO_3 - Nb_2O_5 phase diagram.
c) A variety of microstructures can be controllably produced in $KNbO_3$ films, ranging from fine-grained polycrystalline to epitaxial single crystal.
d) Polycrystalline films can be produced on Si (100) substrates, with an amorphous SiO_2 interface layer always observed.
e) Heteroepitaxial single crystal films can be deposited on MgO (100) substrates at 600°C.
f) Two types of defects can be observed in the epitaxial films, viz antiphase domain boundaries and {211} growth twins. The effect of these defects on properties have not been determined definitively.
g) The epitaxial films appear to have tetragonal symmetry at room temperature, rather than the expected orthorhombic symmetry.

h) Preliminary dielectric measurements have been performed on polycrystalline films on TiN /MgO substrates.

j) A laser beam (λ = 633 nm) has been coupled into epitaxial films on MgO substrates. Several modes or overtones have been observed. Scattering losses are low.

k) Microstructure-property relationships are presently being determined.

ACKNOWLEDGMENTS

The authors would like to thank Drs. P.D. Hren, J.H. Hren, D. Maher, K. Bachmann and R.F. Davis for useful discussions.

This work is supported by the Office of Naval Research under Contract N00014-88-K-0526, while partial support is provided by the Defense Advanced Research Projects Agency, the National Science Foundation and the Department of Energy (A.R.K).

REFERENCES

1. R.L.Holman, L.M.A. Johnson, D.P. Skinner, "The Desirability of Electrooptic Ferroelectric Materials for Guided Wave Optics," In Proceedings of the 6th IEEE Int. Symp on the Appl. of Ferroelectrics, p. 32-40, IEEE New York, 1986.

2. B.A. Tuttle, "Electronic Ceramic Thin Films, Trends in Research and Development," *MRS Bulletin*, Oct/Nov, 40(1987).

3. E.T. Wu, J.J. Ratto, J.R. Oliver, and RR. Neurogaonker, "Growth and Properties of Ferroelectric $KNbO_3$ Single Crystals," In Proceedings of the 6th IEEE Int. Symp on the Appl. of Ferroelectrics, p. 478-81, IEEE New York, 1986.

4. Landolt-Börnstein, Numerical Data and Functional Relationships in Science and Technology, New Series, Vol 16a, Eds. T. Mitsui and S. Nomura, pgs. 48-52, Springer-Verlag, Berlin, 1981.

5. A. Reisman, "Phase Equilibria in the System K_2CO_3 - Nb_2O_5 by the Method of Differential Thermal Analysis," Soc. **77**, 2115 (1955).

6. A.I. Kingon, O.H. Auciello, M.S. Ameen, C.N. Soble, T. Graettinger, S.H. Rou and A.R. Krauss, "Computer-Controlled Ion Beam Sputter Deposition of Multicomponent Oxides," accepted for publication in the Proceedings of the Symposium on Ceramic Thick and Thin Films, Indianapolis, April 1989.

7. S.M. Rossnagel and J.J. Cuomo, "Negative Ion Effects During Magnetron and Ion Beam Sputtering of $YBa_2Cu_3O_7$," Eds. J.M.E. Harper, R.J. Colton and L.C. Feldman, Amer. Inst. of Phys Conf. Proc. No. 165, 106-113 (1988).

8. N. Terada et al, "Sputter Synthesis of YBa_2Cu_3Oy As-Deposited Superconducting Thin Films from Stoichiometric Targets: A Mechanism of Compositional Deviation and its Control," Jap. J. of Appl. Phys. **27**, L639-42 (1988).

9. O. Auciello, A.R. Krauss, A.I. Kingon and M.S. Ameen, "A Critical Analysis of Techniques and Basic Phenomena Related to Deposition of High Temperature Superconducting Thin Films," accepted for publication in SEM Int.

10. O. Auciello, M.S. Ameen, T.M. Graettinger, S.H. Rou and A.I. Kingon, "Ion Beam Sputter Deposited $YBa_2Cu_3O7\text{-}\sigma$: Beam Induced Target Changes and their Effect on Deposited Film Composition," accepted for publication in the Amer. Inst. of Phys. Conf. Proc. (AVS Annual Meeting, Atlanta, Oct. 1988).

11. M.S. Ameen, O. Auciello, A.R. Krauss, A.I. Kingon and M.A. Ray, "Studies on Ion Scattering and Sputtering Processes in Ion Beam Sputter Deposition of High Temperature Superconducting Films," published in AIP Conf. Proceedings, **182**, 61 (1989).

12. M.S. Ameen, O. Auciello, A.R. Krauss and A.I. Kingon, "Studies on Ion Scattering and Sputtering Processes Relevant to Ion Beam Sputter Deposition of Multicomponent Thin films," submitted to the Proc. of the Mat. Res. Soc. Fall Meeting, Boston, Nov. 1989.

13. A.R. Krauss, O. Auciello, A.I. Kingon, M.S. Ameen, T. Barr and Y.L. Liu, "A Novel Computer Controlled Ion Beam Sputter Deposition System," submitted to Rev. Sci Instr.

14. S.H. Rou, P.D. Hren and A.I. Kingon, "Preparation of Cross-sectional and Plan View TEM Samples of Thin Films on MgO Substrates with an Additional Chemical Thinning Step," to be presented at the Mat. Res. Soc. Spring Meeting, San Francisco, April 1990.

15. F.S. Galasso, "Structure and Properties of Inorganic Solids," p.189, Pergamon Press, Oxford (1970).

16. S.H. Rou, P.D. Hren, T.M. Graettinger, M.S. Ameen, O.H. Auciello and A.I. Kingon, "High Resolution Imaging of Twin and Antiphase Domain Boundaries in Perovskite $KNbO_3$ Thin Films," to be presented at the Mat. Res.Soc. Spring Meeting, San Francisco, April 1990.

17. A.I. Kingon, S.H. Rou, P.D. Hren, M.S. Ameen, T.M. Graettinger, K. Gifford and O. Auciello, "Microstructure-Property Relationships for Ferroelectric Thin Films Deposited by Ion Beam Sputtering, to be submitted to the IEEE Transactions on Ultrasonics, Feroelectrics and Frequency Control (Special Issue on Ferroelectric Thin Films) To be published May 1990.

ACOUSTO-, ELECTRO-AND NONLINEAR OPTICS IN ZnO FILMS

T. Shiosaki

Dept. of Electronics, Faculty of Engineering, Kyoto University, Kyoto 606, Japan

ABSTRACT

The developments in the acousto-, electro-, and nonlinear optical interactions in ZnO films on glass, silicon and sapphire substrates are described.

I. INTRODUCTION

Zinc oxide, a member of the hexagonal wurtzite class, has very strong piezoelectric effect among nonferroelectric material. It has large piezoelectric, sizable elasto-optic, electro-optic and nonlinear optic coefficients. It is also transparent from 0.4 to 2 μm due to the 3.3eV wide bandgap. Thin-film ZnO has been used extensively as a piezoelectric transducer for the generation and detection of bulk and surface acoustic waves. The ZnO thin film was usually deposited either beneath or above the electrode array for SAW transducers which can generate surface acoustic waves with power enough to attain high diffraction efficiency in surface acoustooptic devices on nonpiezoelectric substrates. The ZnO film is also very suitable as an optical waveguide and can be used as the interaction medium in surface acoustooptic and electrooptic devices. Thus, using a low optical loss ZnO waveguide, it can perform a variety of active and passive operations on guided light such as have been proposed for integrated optical systems.

The RF planar magnetron sputtering system was used for the present ZnO film growth. The $(01\bar{1}2)$ sapphire, Corning 7059 glass, and p-type (100) Si with 1000-2000 Ω cm were used as substrates of ZnO films for the present acousto-optic deflectors. The Si substrates were covered with a 6-μm-thick layer of sputtered SiO_2. Details of the growth and the characterization of the ZnO films on sapphire are described elsewhere. The ZnO film is one of the suitable materials as an optical waveguide and can be used as the interaction medium in surface acoustooptic devices[1-5]. Since optical scattering loss must be kept to a minimum in waveguides, a continued effort has been made to get low-loss ZnO films for optical applications. Although the optical waveguiding intensity profile of the scattered light from the ZnO films on glass and Si was a little wavy than that on sapphire, the optical loss was also around $0.5 dBcm^{-1}$. The minimum optical loss was less than $0.1 dBcm^{-1}$ for one of the ZnO films on glass.

Diffraction ratios of 95 and 98 percent for the TE_0 mode beam (at 632.8nm) were accomplished by acoustic powers of 90mW in a ZnO film on glass and 80mW in a ZnO film on Si, respectively. The diffraction ratios of the AO Bragg deflectors using a ZnO film on sapphire with IDT's of different Q values and with the surface-acoustic-wave (SAW) propagating along and perpendicular to the c-axis of the ZnO film were also measured[6].

An electrooptic Bragg diffraction modulator with high diffraction ratio was obtained using a low-loss epitaxial ZnO optical waveguide on a sapphire substrate. The unclamped electrooptic coefficient r_{33} is estimated to be 5.8×10^{-12} m/V, which is much larger than the clamped value of 2.6×10^{-12} m/V listed in the literature[7].

A grating-type Schottky-barrier photodetector was integrated with a ZnO optical waveguide on a silicon substrate. Silicon was chosen as the base material for the future integration of optical and electronic components.

SHG in an as-sputtered ZnO optical waveguide on a fused-quartz substrate was performed with a Nd: YAG laser at 1.064 μm wavelength. For 150mW fundamental input power, the observed conversion efficiency was about 4.9×10^{-6}. The phase-matched interaction length of 2.5mm was about 125 times larger than the previous result. The results reveal another potential use of sputtered ZnO film in integrated optical applications[8].

II. SAW ACOUSTOOPTIC BRAGG DIFFRACTION

Experiments on acoustooptic deflectors were made in the Bragg diffraction region of the criterion index $Q \geq 10$. The deflected angle from the direction of the undeflected beam is $2\theta_B$, where θ_B is the external Bragg angle given by $\theta_B = \sin^{-1}(\lambda/2\Lambda)$. The average value of $2\theta_B$ measured agrees with the theoretical value of 1.81^o within our measurement error of $\pm 0.03^o$. The acoustic power was determined from measurements of the RF input power and the transducer insertion loss. Comparing the insertion loss of two SAW delay lines of different propagation lengths, the propagation loss was estimated to be $1.2 dBcm^{-1}$. The laser light passes through the center part between the input and output IDT's. Therefore, the acoustic power is deduced to be a half value of the insertion loss because the propagation loss is negligibly small. Two kinds of diffraction ratios I_1/I_0, η_A and η_B, where η_A is the fractional decrease of the output power in the undeflected optical beam and η_B is the ratio of the output power in the deflected optical beam to that of the nondiffracted optical beam, were measured. Figs. 1 and 2 show η_A and η_B which were given in terms of the square root of acoustic power for SAW propagating along and perpendicular to the c-axis of ZnO film, respectively. Since the SAW propagating along the c-axis

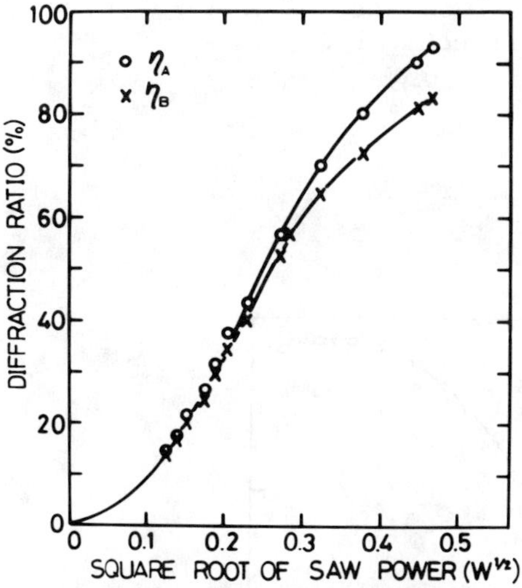

Fig.1

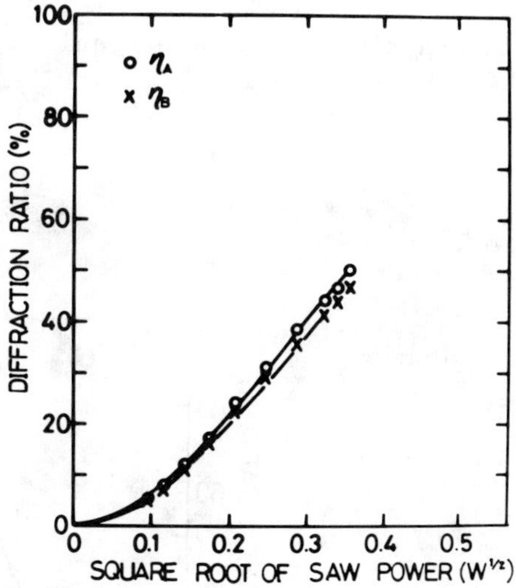

Fig.2

Fig.1 Dependence of diffraction ratios η_A and η_B on square root of acoustic power using an IDT of Type A with SAW propagating along the c-axis of ZnO film with 1.55-μm thickness on sapphire. Acoustic power was 9dB below RF input power. Q of Type A IDT is 10.

Fig.2 Dependence of diffraction ratios η_A and η_B on square root of acoustic power using IDT of Type A with SAW propagating in direction perpendicular to c-axis of ZnO film with 1.55-μm thickness on sapphire. Acoustic power was 11.8dB below RF input power. Q of Type A IDT is 10.

of ZnO film is a Rayleigh-type wave, the strain S_3, through the photoelastic constant p_{33}, makes the dominant contribution to the acoustooptic interaction. On the other hand, the strain S_5, through the photoelastic constant p_{44}, makes the dominant contribution to the acoustooptic interaction when the SAW propagates in the direction perpendicular to the c-axis as a Love-type wave.

A step-like ZnO film optical waveguide on Corning 7059 glass was prepared. An IDT was deposited at the interface between the ZnO film and glass. The thickness of the transducer ports was determined to be 9.6 μm so that the maximum value of the effective coupling factor would be at a center frequency around 130MHz. The insertion loss was measured to be 18.8dB at the center frequency of 124MHz. Fig. 3 shows η_A in terms of the square root of the acoustic power. The maximum diffraction ratio is 95 percent at an acoustic power of about 90mW. Total electric drive powers of 141 and 766mW for 50 and 95 percent diffraction ratios are required, respectively.

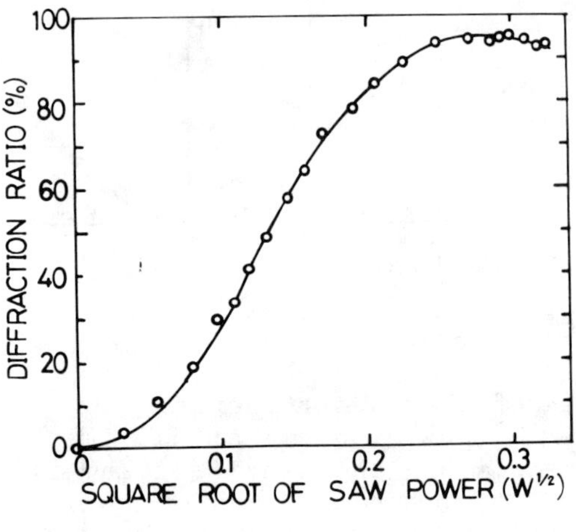

Fig.3 Dependence of diffraction ratio η_A on square root of acoustic power using IDT of Type D fabricated at interface between ZnO film and glass. Acoustic power was 9.3dB below RF input power. Q of Type D IDT is 15.

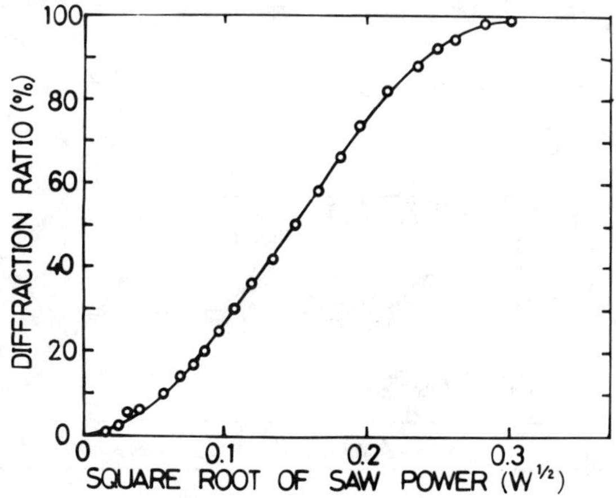

Fig.4 Dependence of the diffraction ratios η_A and η_B on the square root of acoustic power using IDT of Type D fabricated on the ZnO film surface on Si. The acoustic power is -14.5dB under the rf input power. Q of Type D IDT is 15.

The AO deflector on Si substrate was also fabricated because it would allow future integration of optic and electronic components. An IDT was fabricated on a 2.0 μm thick ZnO film. The maximum diffraction ratoio is as high as 98 percent at an acoustic power of about 80mW as shown in Fig.4. It requires a total electric drive power of 585mW for 50 percent diffraction ratio.

All the diffraction ratio measurements were at 632.8nm for the TE_0 mode.

III. ELECTROOPTIC BRAGG DIFFRACTION

Fig.5 shows the schematic diagram of the electrooptic diffraction modulator measurements. An interdigital electrode with a spatial period of 20 μm and an aperture of 3mm of 1000 Å thick Al was deposited directly onto the ZnO waveguide surface using the photoresist lift-off technique. At 6328Å, the criterion index Q is 15, indicating that the diffraction in this experiment is strongly governed by Bragg diffraction.

As interdigital electrode is deposited directly onto the waveguide, the mode characteristics of the waves under the electrode strip and the bare surface are different.

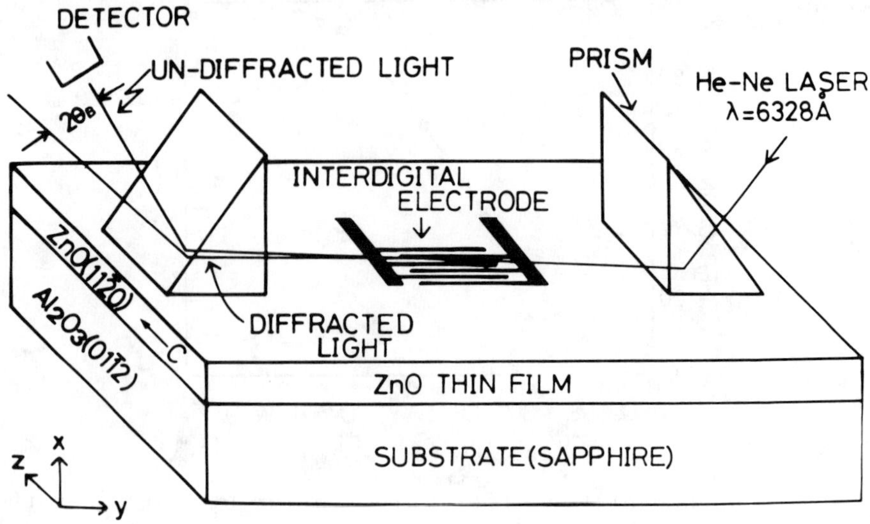

Fig.5 Schematic diagram of the electrooptic grating waveguide modulator.

Consequently, even with no voltage applied to the modulator, the interdigital electrodes would act as a grating and cause the beam to diffract, and its diffraction angle is expected to be twice the diffraction angle due to the applied electric field.[9]

Two kinds of diffraction ratios η_A and η_B are usually used in the experiments. η_A is the fractional decrease in the output from the undiffracted optical beam, and η_B is the ratio of the output of the diffracted optical beam to that of the undiffracted optical beam. The value η_A is more reliable and was used in our measurements.

A dc power supply and sinusoidal wave oscillators with frequencies from 10Hz to 10MHz was used as the input source for the interdigital electrode. All of the measurements were made for the TE_0 mode at 6328Å.

The maximum diffraction ratios were over 90 percent for almost all the samples with thickness from 0.8 μm to 1.2 μm. It is still not very clear how the applied voltage V_f corresponding to the maximum ratio is related to the thickness of the film. But it seems that V_f tends to be higher for those samples with larger thickness.

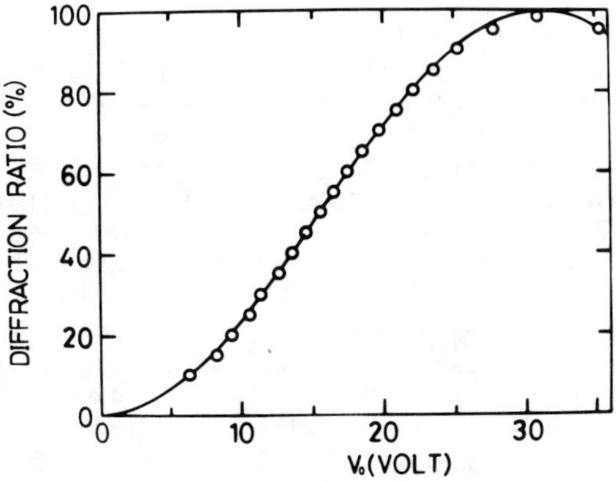

Fig.6 Dependence of the diffraction ratio η_A on applied voltage V_0 of sample C with a 0.94 μm thickness.

Fig.6 shows our best diffraction ratio η_A measurement, of the sample with thickness of 0.94 μm using a dc power supply. The solid curve is $\sin^2 BV_0$ normalized to measured data at η_A=70 percent. The data follow the sine-squared law very well. The maximum diffraction ratio η_A is about 98 percent at an applied voltage of about 31V. The reproducibility was quite good.

The resistivity of ZnO film after sputtering was about 10^6 -10^7 Ω·cm. But it decreased to $10^4 - 10^5$ Ω·cm after four months or half a year and became steady.

The low-frequency (unclamped) electrooptic coefficient r_{33} of ZnO film is calculated from ref.10 to be about 5.8×10^{-12}m/V, which is much larger than the clamped value of 2.6×10^{-12}m/V listed in the literature. There is no listed value of the unclamped electrooptic coefficient r_{33} of ZnO in the literatures. It is very interesting that there is such a large difference between the clamped and the unclamped r_{33} values of ZnO. The results were independent of the frequency in the range from dc to 10MHz.

Since it is very lossy for TM mode for our interdigital electrode fabricated directly on the ZnO film surface, we cannot measure the unclamped electrooptic coefficient r_{13} of ZnO. One solution to this may be to sputter a very thin (∿1500Å) SiO_2 or Al_2O_3 buffer layer between the ZnO thin film and the interdigital electrode.

IV. INTEGRATED SCHOTTKY-BARRIER PHOTODETECTOR

The use of grating-type Schottky-barrier diodes as photodetectors has been reported by Wang and Li[11]. The quantum efficiency of their detector is limited by the fact that the light reflected from the metal electrode is lost to the detector. But by placing the metal grating electrodes at one boundary of the planar waveguide and in the path of the guided wave, the light reflected from the metal electrodes is kept in the waveguide. Thus it will not be lost but eventually be absorbed by Si. Because the reflectance is considerably higher at the metal surface than at the Si surface, most of the guided light is absorbed in Si and not in the metal. It is expected, therefore, that with the grating-type Schottky-barrier diode, the detection of a guided wave can be achieved with excellent sensitivity.

In the present study, a grating-type Schottky-barrier photodetector was integrated with a ZnO optical waveguide on a silicon substrate. In the configuration of grating electrode, d_1 is the width of the electrode, d_2 is the spacing between the electrodes. w is the total width of the grating electrode along the direction of light propagation and l is the length of the electrode. The guided optical light can convert itself into electrical energy more efficiently if the width of the electrode d_1 is smaller and the spacing between the electrode is larger while keeping the depletion layer from breaking. We chose d_2 at 20 μm and d_1 at 5 μm so that the ratio between the width of Pt and Si is one to four. Then the average attenuation constant α_{av}, is $\alpha_{av}=\alpha_{Pt} \times 1/5 + \alpha_{Si} \times 4/5=177.5\text{cm}^{-1}$. The total width w of the grating electrode for absorption of 99% guided light can be decided according to the equation.

Silicon n-type (100) wafers of resistivity 100-120Ω·cm with a 450 μm thickness were used as the base material in this experiment. We cannot form a ZnO optical waveguide directly on Si since the refractive index of Si is larger than the refractive index of ZnO. An isolating layer of low-index dielectric film is needed. This can be achieved by fabricating a SiO_2 layer. The thickness of SiO_2 film should be at least larger 0.4 μm to obtain a low-loss waveguide structure.

For evaluation of Schottky-barrier photodetector, a TE_0 mode He-Ne laser light of wavelength 6328Å was focused and coupled into the ZnO optical waveguide through a rutile prism. A chopper was placed in the path of the input laser beam. The photodetector was reversely biased at V_R=8V through a load resistance R_L=30KΩ. The output voltage across the load resistance was connected to an oscilloscope. Fig.7(a) shows the display of the reference signal of the optical chopper and

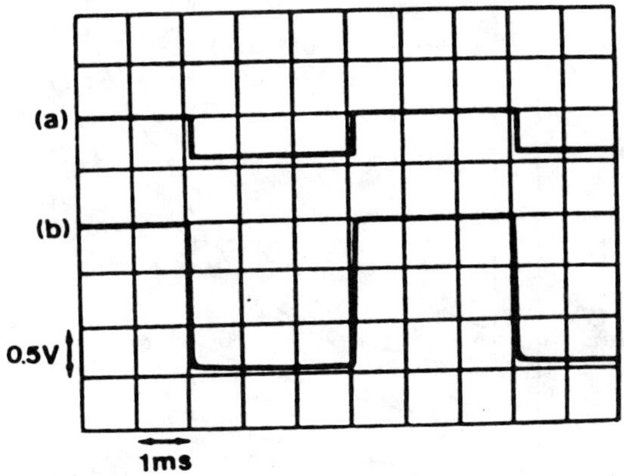

Fig.7 (a) Reference signal of optical chopper. (b) Detected signal by Schottky-barrier photodiode. Vertical scale: 0.5V/div.: horizontal scale: 1 ms/div..

Fig.7(b) shows the display of the detected signal. The frequency of the chopper was 100Hz. The rise up time is about 15 μs. To improve the response time of the detector, we should try to reduce the diode capacitance and series resistance. We don't need 100% absorption of light if we only need to detect the optical signal.

The experiment of the detection of the AO and EO deflected guided wave on Si substrate is now under way.

V. SECOND HARMONIC GENERATION

The refractive indices for the ZnO film were determined, using the m-line method described by Tien et al.[12] as n_o=1.942 and n_e=1.957 at 1.064 μm and n_o=2.029 and n_e=2.046 at 0.532 μm. On the other hand, because of the lack of precise refractive index data for fused quartz, the Sellmeier dispersion formula was tentatively employed to estimate the values for the substrate, giving n=1.4497 at 1.064 μm and 1.4608 at 0.532 μm. The modal dispersion curves calculated from these refractive indices are shown in Fig.8. In the following SHG experiments, conversion from $TE_0^{\omega} + TE_0^{\omega}$ to $TM_2^{2\omega}$ was attempted in order to utilize the highest spatial coupling coefficient existing between these modes. From Fig.8, the film thickness required for the complete phase matching is determined as 0.970 μm for this interaction.

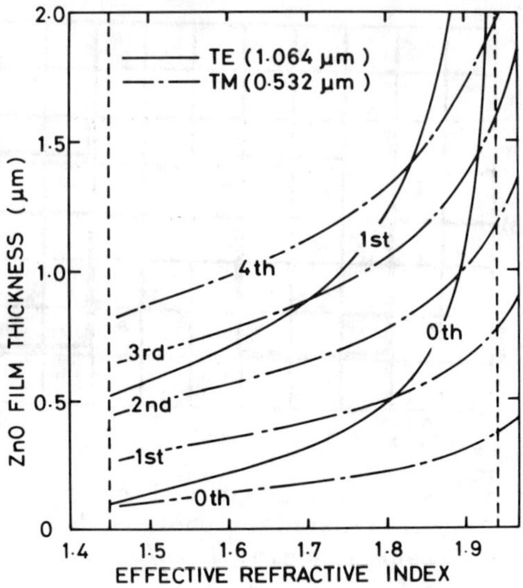

Fig.8 Modal dispersion curves calculated for the sputtered ZnO film.

The SHG experiments were performed using a linearly-polarized 1.064 μm Nd: YAG laser as the fundamental light source. The guided TE modes were excited by focusing the incident beam onto a rutile prism coupler with a lens of 10.0cm focal length. The guided modes were identified by measuring the incident angle to the prism coupler as well as by displaying the m-lines on an Si-vidicon TV monitor. The second harmonic output was separated from the fundamental with an infrared-cut filter and a polarizer. A thermopile optical power meter and calibrated photomultiplier were used to determine the output intensity.

A second-harmonic bright green streak was observed on the film surface for the Q-switched excitation of the fundamental TE_0 mode when complete phase matching was satisfied by sliding the tapered-film waveguide across the incident beam. Figure 9 shows the polarization state of the second harmonic output as a function of the rotation angle of the polarizer, indicating that the output green light is in a TM mode. Besides, the output emerging angle from the output prism coupler was consistent with the expected TM_2 mode. Figure 10 shows a spectrum trace of the output light. The peak wavelength is located at 0.532 μm and corresponds to the second harmonic of the input wavelength.

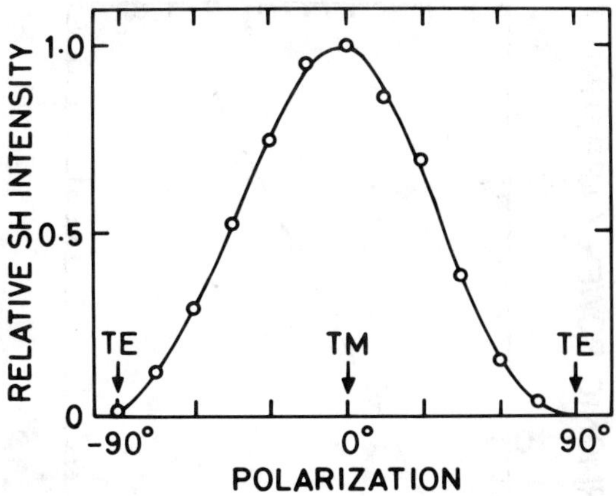

Fig.9 Polarization state of the second harmonic output as a function of rotation angle of the polarizer.

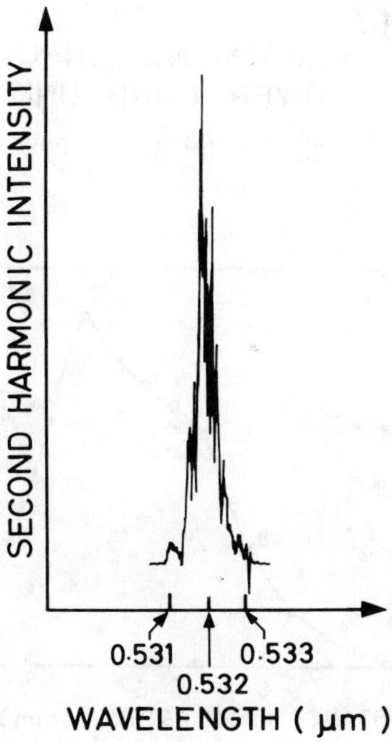

Fig.10 Spectrum trace of the output second harmonic radiation. Resolution 8Å.

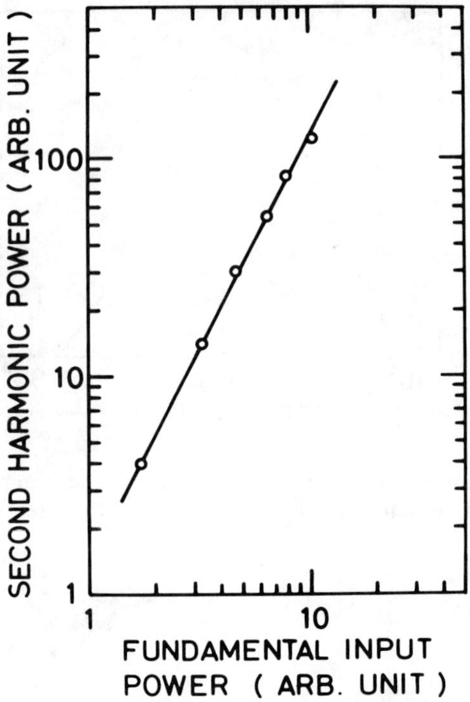

Fig.11 Dependence of the second harmonic power on the fundamental input power.

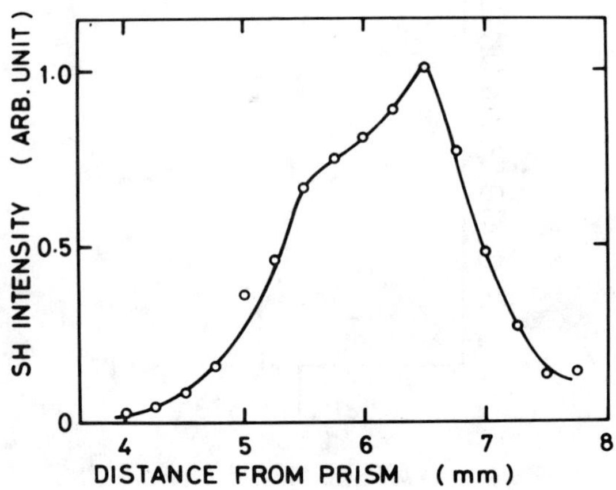

Fig.12 Variation of the second harmonic intensity with propagation distance.

Figure 11 shows the dependence of the second harmonic output power on the input power. The quadratic dependence is in agreement with the theoretical prediction. Figure 12 shows the growth of the second harmonic radiation with propagation distance measured by an optical fiber probe. The second harmonic starts to grow at about 4mm from the input prism edge and markedly increases to the position of about 6.5mm, indicating that constructive interaction took place over a distance of at least 2.5mm in this geometry. The allowable tolerance for the film thickness to obtain the phase matching for a given interaction length l can be calculated by employing a technique similar to that described by Suematsu et al.[13]. From the calculation assuming l=2.5mm, the deviation of the film thickness from the precisely phase-matchable value is estimated to be less than 10Å over the entire distance of 2.5mm in the present sample.

For 150mW cm fundamental input power, the second harmonic power obtained at the position 6.5mm from the prism edge is 7.3 x 10^{-7}W, and the conversion efficiency is therefore 4.9 x 10^{-6}. The effective nonlinear optical coefficient giving rise to this SHG is considered to be d_{31}^{2}[11].

CONCLUSION

1) The quality of ZnO films was improved.
2) The diffraction ratio of 98% in the guided wave SAW-AO Bragg diffraction modulator was attained.
3) The distinction ratio of 20:1 was attained in the SAW-EO Bragg diffraction modulator.
4) The optical second harmonic green guided waves were generated from the YAG laser beam guided in the ZnO film.
5) The guided waves in the ZnO film on a silicon substrate were detected by the Schottky diode integrated on the same silicon substrate as the ZnO.

REFERENCES

[1] P.K. Tien, R. Ulrich, and R.T. Martin, "Modes of propagating light waves in thin deposited semiconductor films", Appl. Phys. Lett., Vol.14, pp.291-294, May 1969.

[2] J.M. Hammer, D.J. Channin, M.T. Duffy and J.P. Wittkle, "Low-loss epitaxial ZnO optical waveguides", Appl. Phys. Lett., Vol.21, pp.358-360, Oct. 1972.

[3]K. Setsune, T. Mitsuyu, and K. Wasa, " High frequency AO Bragg deflector using a single-crystal ZnO thin film," in Proc. 3rd Symp. Ultrason. Electron., Tokyo, Japan, Jpn. J. Appl. Phys., Vol.22, Suppl. 22-3, pp.173-175, 1983.

[4]T. Shiosaki, S. Ohnishi, Y. Murakami, and A. Kawabata, "Higher rate epitaxial growth of ZnO films on sapphire by planar magnetron RF sputtering system," J. Cryst. Growth, Vol.45, pp.346-349, Dec. 1978.

[5]T. Shiosaki, S. Ohnishi, and A. Kawabata, " Optical properties of single-crystalline ZnO film smoothly chemical-vapor deposited on intermediately sputtered thin film on sapphire," J. Appl. Phys., Vol.50, pp.3113-3117, May 1979.

[6]M.S. Wu, A. Azuma, T. Shiosaki and A. Kawabata, "Low-loss ZnO optical waveguides for SAW-AO applications", IEEE Transactions on Ultrasonics, Ferroelectrics and Frequency Control, Vol.36, pp.442-445, No.4, July 1989.

[7]M.S. Wu, T. Shiosaki and A. Kawabata, "Electrooptic diffraction modulation in ZnO film on sapphire", IEEE J. of Quantum Electronics, Vol.25, pp.252-256, No.3, Mar. 1989.

[8]T. Shiosaki, S. Fukuda, K. Sakai, H. Kuroda and A. Kawabata, "Second Harmonic Generation in as-sputtered ZnO optical waveguide", Jpn. J. Appl. Phys., Vol.19, pp.2391-2394, No.12, Dec. 1980.

[9]J.M. Hammer, D.J. Channin, M.T. Duffy and C.C. Neil, "High-speed electrooptic waveguide grating modulator using epitaxial ZnO", IEEE J. Quantum Electron., Vol.QE-11, pp.138-148, Apr. 1975.

[10]Y.K. Lee and S. Wang, "Electrooptic Bragg-deflection modulators:Theoretical and experimental studies", Appl. Opt., Vol.15, pp.1565-1572, June 1976.

[11]C.T. Wang and S.S. Li, "A New Grating-Type Gold-N-Type Silicon Schottky-Barrier Photodiode", IEEE Trans. Electron Devices, Vol.ED-20, pp.522-526, June 1973.

[12]P.K. Tien, R. Ulrich, and R.J. Martin, "Modes of Propagating Light Waves in Thin Deposited Semiconductor Films", Appl. Phys. Lett. Vol.14, pp.291-294.

[13]Y. Suematsu, Y. Sasaki, K. Furuya, K. Shibata and S. Ibukuro: "Optical second-harmonic generation due to guided-wave structure consisting of quartz and glass film", IEEE J. Quantum Electron. QE-10 (1974) 222.

[14]N. Uesugi, H. Ito and H. Inaba, "Optical harmonic generation in oriented polycrystalline nonlinear optical waveguide", Pap. Tech. Group Optical and Quantum Electronics of IECE of Japan, OQE 73-75, 1974.

FERROELECTRIC $Sr_{1-x}Ba_xNb_2O_6$ (SBN) THIN FILM DERIVED BY SOL-GEL TECHNIQUE

Ching Jih Chen, Yuhuan Xu, Ren Xu and John D. Mackenzie
Materials Science and Engineering Department
University of California, Los Angeles, CA 90024

ABSTRACT

Ferroelectric strontium-barium niobate thin films with composition of $Sr_{1-x}Ba_xNb_2O_6$ (x= 0.4), have been fabricated successfully. The highly dense, transparent and crack-free films on silicon and fused silica substrates were obtained. Ferroelectric properties were confirmed by P-E hysteresis loops. The optical property, electro-optic effect, and pyroelectric coefficient were reported.

INTRODUCTION

Ferroelectric crystals with tungsten bronze structure have received considerable attention for twenty years, with the well known of these being composition in the ferroelectric $Sr_{1-x}Ba_xNb_2O_6$(SBN) solid solution system. Recently, increasing interests have been focused to thin film electro-optic materials for making integrated optic devices [1]. Numerous other applications of SBN, such as pyroelectric infrared detectors [2], piezoelectric transducers [3], and optic phase conjugators [4], have also been realized. For such applications, large electro-optic coefficients and optical transparency of thin films are necessary. In addition, materials with controllable stoichiometry and oriented crystallinity are highly desirable for device performance.

A number of techniques to make ferroelectric materials, such as crystal growth by Czochralski method [5], sputtering [6], have been studied. Sol-gel method is relatively new ceramic processing technique. It has been successfully used for preparation of ceramics, glasses, fibers, and thin films . The advantages of the sol-gel method are high homogeneity, low processing temperature, and easy fabrication of large area thin films.

Currently, we have successfully fabricated SBN thin films on silicon and fused silica substrates using the sol-gel method. In this paper, the preparation of SBN thin films by sol-gel method was described. Thin-film properties are reported including the ferroelectric properties, pyroelectric coefficient, optical transmission spectrum, electro-optics effect and microstructure of thin films.

MATERIALS PREPARATION

Figure 1 illustrates the flow diagram for the preparation of solutions and films using a sol-gel method. The choice of precursor compounds and solvents based on the following requirements: the compounds should have high solubility in the selected solvents, and should be chemically compatible with each other. The

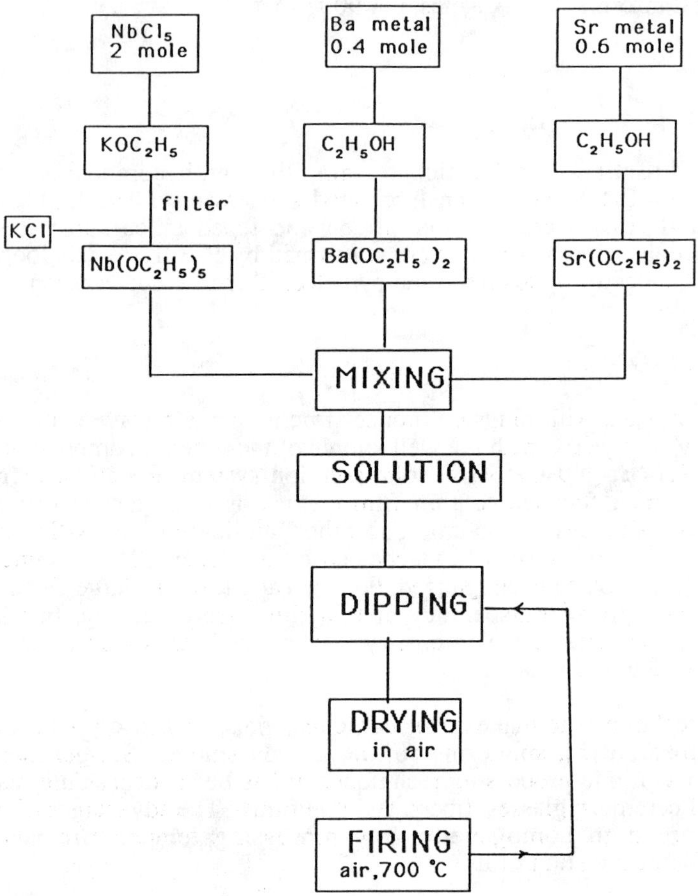

Figure 1. Flow Chart of preparation of SBN film by sol-gel processing.

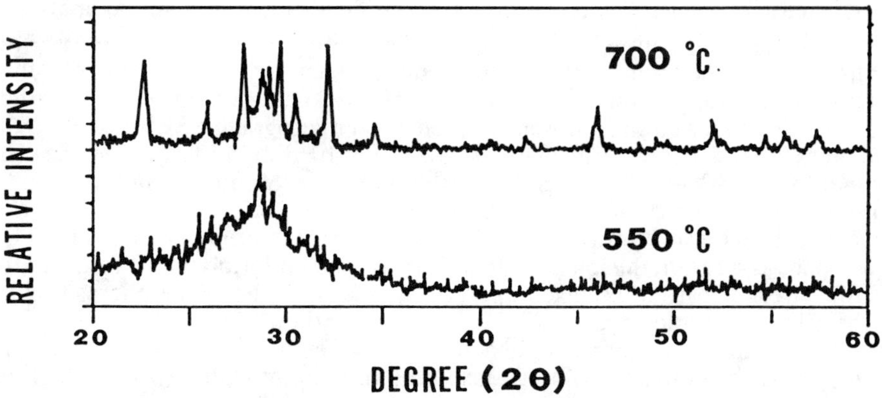

Figure 2. X-ray diffraction patterns of $Sr_{0.6}Ba_{0.4}Nb_2O_6$ films on silicon substrate after heat treatment (a) 550°C, 1 hr, (b) 700°C, 1 hr.

solvent should have the proper boiling point, viscosity, and surface tension. The solution formed should have a high concentration of necessary components and be stable under ambient condition. Based on those considerations, three starting materials, $Ba(OC_2H_5)_2$, $Sr(OC_2H_5)_2$, and $Nb(OC_2H_5)_5$ were used. Solution of $Ba(OC_2H_5)_2$ and $Sr(OC_2H_5)_2$ in ethanol were prepared by direct reaction of these metals with dehydrated ethanol under dry atmosphere. Commercial $NbCl_5$ was used to prepare the solution of $Nb(OC_2H_5)_5$. Potassium metal alkoxide solution in C_2H_5OH was reacted with $NbCl_5$ in dry atmosphere; KCl was then separated as precipitates. Subsequently, the solutions of $Ba(OC_2H_5)_2$, $Sr(OC_2H_5)_2$ and $Nb(OC_2H_5)_5$ were mixed and refluxed for 6 hours and further concentrate to adjust viscosity. The homogeneous alkoxide solution of SBN was used for dip-coating or spin-coating on single crystal silicon and fused silica substrates, and then exposed to wet air for hydrolysis and subsequent polymerization. Prior to coating, the substrates of silicon and fused silica were cleaned with HF (hydrogen fluoride) solution and ethanol respectively. The deposited films were dried at room temperature for 1 hour, then heated up to 700 °C with heating rate of 1°C/min, and hold for 1 hour, finally cooled down to room temperature slowly.

FILM CHARACTERIZATION

The structure of films at different firing temperatures were determined by the x-ray diffraction technique. The film with composition of $Sr_{0.6}Ba_{0.4}Nb_2O_6$ has tungsten bronze structure. Figure 2 shows x-ray diffraction patterns of the films on silicon substrate as a function of firing temperature. Films fired at 550°C for 1 hour, were partially crystallized. As the firing temperature was raised to 700°C for 1 hour, tetragonal phases was observed. From the differential thermal analysis (DTA) curves, the initial crystallization temperature was 590°C. To improve the quality of film, the heat-treatment temperature is usually higher by 100°C than required to get fully crystalline structure. The thickness of films on fused silica substrate were measured from scanning electron micrograph (SEM), where the thickness of 0.8 μm of film was obtained by 10 dipping-firing cycles. The resultant film is dense and crack-free. Although, a large internal stress could be caused by the mismatch of thermal expansion coefficient between film and substrate, the adhesion between the film and substrate is strong enough due to M-O-M bond, as a result, the internal stress is dispersed uniformly in the film and the internal stress is partially relaxed upon firing. Figure 3 shows the microstructure of the film on single crystal silicon substrate without poling. The individual crystalline sample contains multiple domains of different ferroelectric orientations. After poling, the domains inside grains in the film orient with respect to the field direction so that the ferroelectric films become anisotropic symmetry corresponding to the macroscope symmetry of point group ∞**m.**

Figure 3. The microstructure of SBN (doped with 0.3% Cu) film on single crystal silicon substrate, where the ferroelectric domains were observed.

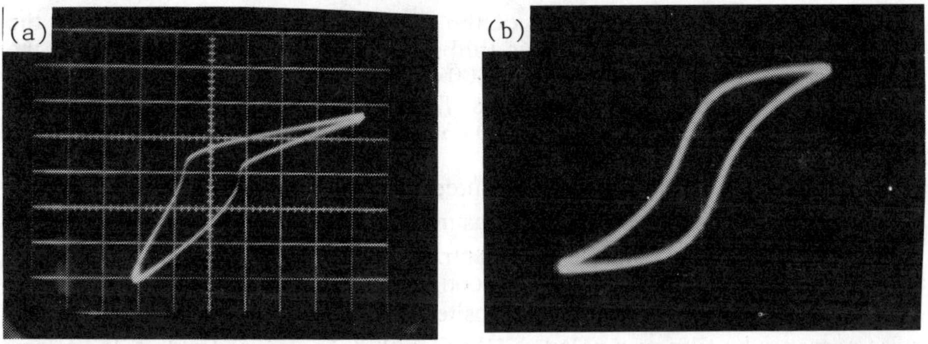

Figure 4. (a) The asymmetric P-E hysteresis loop of SBN on p-Si substrate, the remanent polarization P_r is 18 $\mu C/cm^2$, and the coercive field E_c is 68 kV/cm, scale of x axis of loop field E is 86 kV/cm/div, scale of y axis of loop polarization P is 17 $\mu C/cm^2$/div. (b) the symmetric P-E hysteresis loop of the film on fused silica substrate.

PHYSICAL PROPERTIES

To measure the ferroelectric properties, the gold electrodes were deposited on the SBN films by sputtering . Ferroelectric hysteresis loops measurements were made using a modified Sawyer-Tower circuit at a frequency of 50 Hz. The dielectric loss was compensated in each measurement. Figure 4 shows the ferroelectric hysteresis loop observed in the $Sr_{0.6}Ba_{0.4}Nb_2O_6$ films on p-Si (electrode arrangement with Au/SBN/Si sandwich configuration) and fused silica (eleltrode arrangement with Au/SBN/Au parallel configuration) substrates respectively. The silicon (p-type, 111 plane) has resistivity of 1 ohm-cm at room temperature. The remanent polarization P_r and the coercive field E_c measured from SBN/p-Si are 18 $\mu C/cm^2$ and 68 kV/cm respectively. Obviously, the coercive field is larger than that of single crystal with the same composition [5]. The possible reason for this is that the majority of the external field is applied on the ferroelectric/semiconductor junction region, and only a small portion is actually applied on the film. Consequently, the data E_c measured is not the real coercive field of SBN film. It is difficult to obtain data for the SBN film on fused silica, because of the uncertainty of the electrode area. The ferroelectric hysteresis loop of SBN film on fused silica substrate is symmetric, but the asymmetric hysteresis loop of SBN film on Si substrate was observed due to the heterojunction effect [7].

The films deposited on fused silica were transparent. The optical transmission of the films were measured in the wavelength range of 0.2 μm to 1.2 μm by Perkin Elmer 330. Figure 5 shows the optical transmission spectrum of SBN film on fused silica. The refractive index of the film is 2.31± 0.005 measured at the wavelength of 6328 Å by Automatic Ellipsometer (Auto E1-1I). To measure the electro-optic effect, external power of 300 to 1200 volts were applied to the films with width of 2 mm between electrodes. The refractive index changed, when the external applied power increased from 0 volts to 1200 volts.

The pyroelectric coefficient was measured using the dynamic technique [2], in which the samples were heated and measured in the temperature range of 20°C to 140°C using hot plate, and the pyroelectric current i_p was measured by a pico-ammeter (417 Keithley Instruments) connected in series to the sample. A gold electrode with area of 4.0 cm^2 was deposited on SBN films by sputtering. Prior to measurement, the film was poled at 100°C with d. c. electric field of 25 kV/cm. Since

$$i_p/A=(dP/dT)(dT/dt),$$

where A is the area of electrode, and the temperature gradient dT/dt and pyroelectric current i_p were measured, the pyroelectric coefficient p (= $|dP/dT|$) can be determined (Figure 6). The pyroelectric coefficient is about $2x10^{-8}$ C/cm ^{2} K at room temperature, which is in the same order of magnitude of that of SBN single crystal [5]. Figure 6 shows the diffuse peak (like a hill) in the curve of pyroelectric coefficient p(T) in the vicinity of the Curie point T_C. A possible

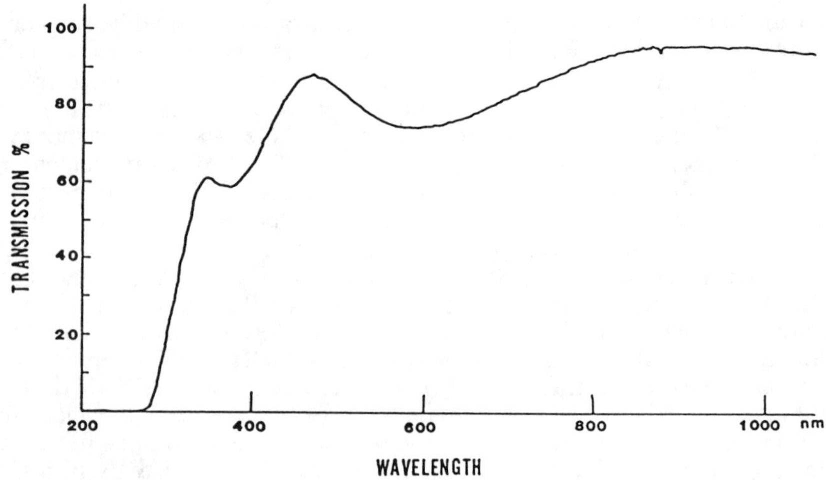

Figure 5. Optical transmission as a function of wavelength for the SBN film with thickness of 0.3 μm.

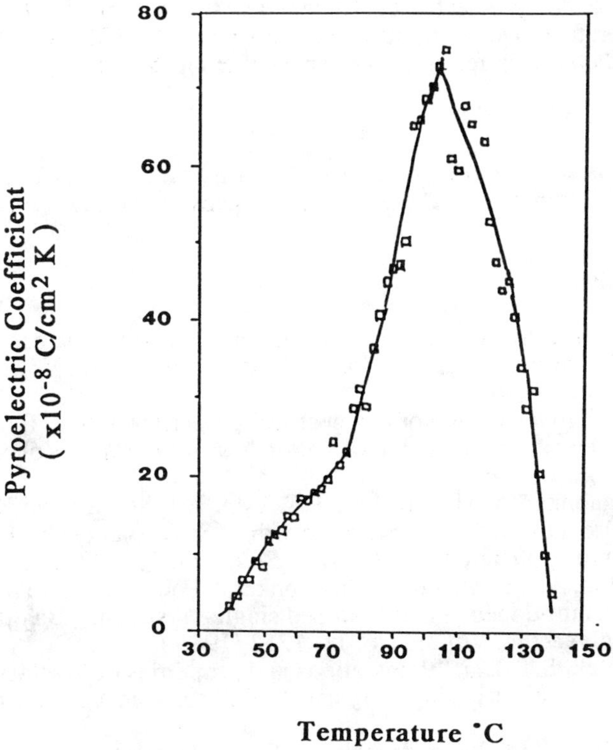

Figure 6. The pyroelectric coefficients of SBN film measured at various temperature.

explanation is that there is a fluctuation of composition in some solid solutions of ferroelectrics with perovskite structure or tungsten-bronze structure, such as SBN, having diffuse phase transition (DPT)[8]. Therefore, the fluctuction of Curie temperature T_C causes the peak of the curve of pyroelectric coefficient p(T) to diffuse.

CONCLUSION

Polycrystalline SBN thin films were prepared by dip-coating solution of complex alkoxides. It is possible to grow a crack-free film on silicon and fused silica substrates. The refractive index (n=2.31) of films are comparable to that of single crystal indicating the film is high density. Also, the refractive index is function of applied electric field. The pyroelectric coefficient as a function of temperature was reported, where the pyroelectric coefficient was found to be about $2x10^{-8}C/cm^2K$ at room temperature. This suggests that the grains on the SBN polycrystalline film by sol-gel method are preferrentially oriented after electric poling.

ACKNOWLEDGEMENT

This work was supported by the Air Force Office of Scientific Research, Directorate of Chemical and Atmosphere Sciences under Grant No. AFOSR-88-0066.

REFERENCES

1. P.V. Lenzo, E.G. Spencer and A.A. Ballman, " Electro-optic coefficients of ferroelectric strontium barium niobate ", *Appl. Phys. Lett.*, Vol.**11**, No. 1, 23 (1967).
2. A.M. Glass, " Investigation of the electrical properties of $Sr_{1-x}Ba_xNb_2O_6$ with special reference to pyroelectric detection ", *J. Appl. Phys.* Vol.**40**, No. 12, 4699 (1969).
3. R.R. Neurgaonkar, M.H. Kalisher, T.C. Lim, E.J. Staples and K.L. Keester," Czochralski single crystal growth of $Sr_{0.61}Ba_{0.39}Nb_2O_6$ for surface acoustic wave applications ", *Mat. Res. Bull.* Vol. **15**, 1235 (1980).
4. G.A. Rakuljic, A. Yariv and R. Neurgaonkar, " Photorefractive properties of undoped, cerium-doped, and iron-doped single-crystal $Sr_{0.6}Ba_{0.4}Nb_2O_6$ ", *Optical Engineering*, Vol.**25**, No. 11, 1212 (1986).
5. R.B. Maciolek, S.T. Liu, "Preparations and properties of low loss $Sr_{1-x}Ba_xNb_2O_6$ ferroelectric single crystals ", *J. Elec. Mat.* Vol. **2**, No.2, 191 (1973).
6. R.N. Castellano, " Ion beam deposition of ferroelectric thin films sputtered from multicomponent targets ", *J. Vac. Sci. Technol.*, Vol. **17**, 629 (1980).
7. Y.H. Xu, C.J. Chen, R. Xu and J.D. Mackenzie, " Self biased hetero-junction effect of ferroelectric thin film on silicon " *J. Appl. Phys.*, Mar. 15, (1990), in printing.
8. Y.H. Xu, H. Wang, " Microscopic analysis of ferroelectrics with diffuse phase transition ", *J. Appl. Phys. Japan*, Vol. **24 Suppl**. 236 (1985).

IMPLICATIONS OF LOW FILM THERMAL CONDUCTIVITY FOR THE LASER DAMAGE RESISTANCE OF OPTICAL THIN FILMS

John C. Lambropoulos and S.-S. Hwang
Department of Mechanical Engineering
University of Rochester
Rochester, NY 14627

ABSTRACT

Recent measurements of the thermal conductivity of thin films of dielectric materials (oxides and fluorides) have shown that the value for thin films (with thickness in the micron and submicron range) may be as much as two orders of magnitude lower than the value for the corresponding bulk solid. We examine the implications of such low values on the laser damage resistance of thin optical films in view of a micromechanics model consisting of an absorbing inclusion embedded within a non-absorbing thin film which is attached to a substrate. Finite element techniques are used to calculate the temperature in the inclusion, in the film's free surface, and along the film/substrate interface.

1. INTRODUCTION

The resistance to laser damage in optical thin films is an important design consideration towards the development of powerful solid-state lasers. Many experimental data on the critical energy density per unit area have been reported by Walker et al. [1] for nine dielectric films as a function of laser pulse length (5 and 15 ns), wavelength (1.06, 0.53, 0.35, and 0.26 μm), and film thickness (1/8 to 2 wavelengths). These data are in the range 1-40 J/cm^2 for the oxide and fluoride films tested [1]. Experimental results on laser damage for 1 ns pulses have been reported by Lowdermilk and Milam [2] for surfaces of optically polished glass and thin optical films in addition to observations of laser damage in such films. Austin et al. [3] have shown that the film's internal stress also influences the laser damage resistance of the film: They reported that the laser damage resistance increases by a factor of 2-3 when the film's internal stress vanishes, and that either tensile or compressive internal stress lead to diminished laser damage resistance. Walker et al. [4] have discussed the mechanisms that lead to laser damage in dielectric materials (avalanche ionization, multiphoton absorption, and impurity-initiated damage) and have concluded that the impurity model appears more likely.

According to the impurity model, which has been discussed by Hopper and Uhlmann [5] for bulk materials, an absorbing inclusion within a non-absorbing infinite matrix is heated due to the absorption of the incident radiation. The high temperature within the inclusion leads to failure when the temperature reaches some critical value [5]. The model uses the solution by Goldenberg and Tranter [6] who considered the time-dependent heat conduction due to the absorption of radiation Q (power per unit volume) within the inclusion of radius R embedded within the non-absorbing infinite matrix. Due to the dependence of Q on the radius R of the inclusion (which absorbs according to the cross section R^2, so that Q varies like 1/R) the laser damage resistance is minimum when the size of the absorbing inclusion is of order 0.2 μm [4].

Further analysis of the absorbing inclusion within the infinite, non-absorbing matrix has been carried out by Lange et al. [7,8] who showed that this model leads to an expression for the laser damage resistance of the form

$$E_0 = \frac{16}{\pi} T_c (\rho\ c\ k\ t)^{1/2} \tag{1}$$

where E_0 is the value of the critical energy density per unit area, T_c is the critical temperature at failure (typically the melting point of the host), and ρ, c, k are the density, heat capacity, and thermal conductivity of the host material (i.e. the matrix) and t is the duration of the laser pulse. In the work by Lange et al. [7,8] bulk values were used for the thin film thermal conductivity.

Although the density and heat capacity of thin films are close to the values for the bulk materials, this is not the case for the thermal conductivity of thin films. Guenther and McIver [9] have recently discussed the measurement and the effect of the thermal conductivity of thin films in view of recent experimental measurements which show that the thermal conductivity of thin films can be considerably lower than the thermal conductivity of the corresponding bulk solid. The discussion by Guenther and McIver [9] is based on eqn. (1) once the host thermal conductivity is identified with the value for the film. Thus, the question is raised of whether or not eqn. (1) is applicable to thin films of very low thermal conductivity, in view of the fact that eqn. (1) has been derived for an inclusion within an infinite matrix, so that other effects such as the proximity of the inclusion to the free surface, the proximity to the film/substrate interface, the size of the inclusion, or the film thickness do not enter the model.

Recent work by Lambropoulos et al. [10] at the Laboratory for Laser Energetics at the University of Rochester has used a modified thermal comparator method to measure in-situ the thermal conductivity of a large variety of optical thin films (oxides and fluorides) deposited on Si or sapphire substrates. The method is quick and non-destructive. It is based on the work of Powell [11], and operates on the principle that when a heated tip touches a cooler solid, the temperature at the tip soon reaches a steady state value which depends on the thermal conductivity of the solid in contact with the tip. The method also allows the determination of the interfacial thermal resistance R_{int}, defined as the temperature drop per unit power flux across the film/substrate interface. Typical values for the thermal conductivity k_F of submicron thin films are in the range 0.04 to 1 W/m/K for oxide films, and in the range 0.08 to 0.75 W/m/K for fluoride films. Some of these values are two orders of magnitude lower than the corresponding bulk values for the same material. Typical values of the interfacial thermal resistance are in the range 0 to 2.7 K/(W/mm^2).

It is the objective of this report to discuss some of the implications of the low values for the film thermal conductivity on the laser damage resistance of such films, and in particular to discuss the applicability of the scaling expressed in eqn. (1) for film/substrate assemblies involving submicron thin films of low thermal conductivity. We will show that when the film thermal conductivity is low, the model of Goldenberg and Tranter [6] which treats the host as infinite does not properly account for the low film thermal conductivity and for the proximity of the inclusion to the substrate. Thus, the predictions of such a model, encapsulated in the failure criterion of eqn. (1), must be viewed with caution, especially in comparing such predictions to experimental data.

2. MODEL OF ABSORBING INCLUSION EMBEDDED WITHIN FILM

Consider a spherical inclusion of radius R embedded within a film of thickness H, such that the center of the inclusion is at adistance D below the free surface of the film which is assumed to be insulated. The film is supported on a semi-infinite substrate as shown in Fig. 1. The subscripts "F", "I", and "S" identify the film, inclusion, and substrate, respectively. At time $t \geq 0$ the inclusion absorbs power at the rate Q per unit volume, whereas the film and substrate do not absorb any incident radiation. The temperature and the heat flux are assumed to be continuous at all interfaces and to decay to zero sufficiently far from the inclusion.

For the case of the inclusion embedded within an infinite matrix, Goldenberg and Tranter [6] have shown that the temperature distribution has a sigmoidal time dependence, being low at small times and achieving its maximum value at steady state ($t \rightarrow \infty$). The steady state solution is given by (here we identify the matrix or host material with the film material)

$$T = \frac{R^2 Q}{k_I}\left\{\frac{1}{3}\frac{k_I}{k_F} + \frac{1}{6}\left(1 - \frac{r^2}{R^2}\right)\right\}, \quad r < R$$
$$T = \frac{1}{3}\frac{R^2 Q}{k_F}\frac{R}{r}, \quad r > R \tag{2}$$

where r is the spherical radial coordinate measuring from the center of the absorbing sphere. It is clear that such a model cannot account for the proximity of the inclusion to a free surface or to a film/substrate interface, nor can it account for a film/substrate geometry where the inclusion is within an insulating film which is deposited on a conducting substrate.

To account for these effects, we assumed R=H/5 (since inclusions of 0.2 micron size give the least damage resistance and typical film thickness is 1 micron), and we used finite elements to solve the time-dependent heat conduction equations

$$k_I \nabla^2 T_I + Q = (\rho c)_I \frac{\partial T_I}{\partial t} , \quad r < R$$

$$k_F \nabla^2 T_F = (\rho c)_F \frac{\partial T_F}{\partial t} , \quad r > R, -(H-D) < z < D \qquad (3)$$

$$k_S \nabla^2 T_S = (\rho c)_S \frac{\partial T_S}{\partial t} , \quad z < -(H-D)$$

where z is the axial coordinate in a cylindrical system with origin in the center of the inclusion, and ∇^2 is the axisymmetric Laplacian operator in the cylindrical coordinates r_1 and z. Note that $r^2=r_1{}^2+z^2$. The initial condition for the temperature is $T(r_1, z, t=0)=0$. As boundary conditions we assume that the film's free surface at z=D is insulated, and that the temperature and heat flux is continuous at the surface of the inclusion and at the film/substrate interface. For the finite element solution we also assumed that the ratio of the thermal diffusivities $d=k/(\rho c)$ is equal to the ratio of the thermal conductivities.

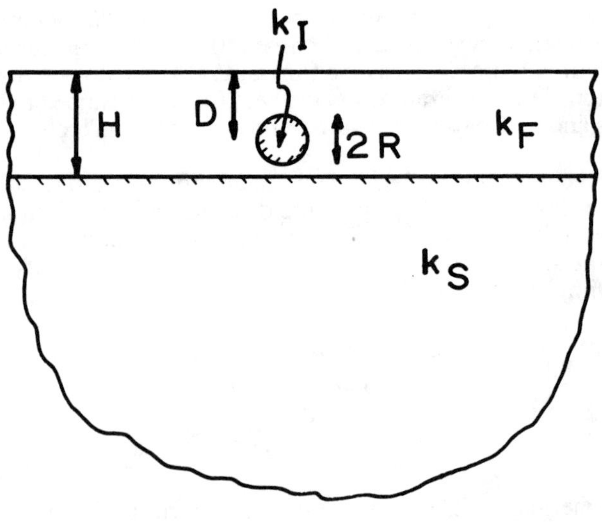

Figure 1. The geometry of a film of thickness H containing an absorbing spherical inclusion of radius R. A cylindrical coordinate system has origin in the center of the inclusion. The film's free surface is located at z=D (assumed insulated). The film/substrate interface is located at z=-(H-D).

The numerical solution showed that the time dependence of the temperature distribution still has the characteristic sigmoidal shape and that the steady state gives the largest temperature values. Figure 2 shows the variation of the steady state temperature (measured in units of $QR^2/2k_S$) at the center of the inclusion vs. the nondimensional ratio of film to substrate thermal conductivity for several values of the ratio of inclusion to substrate thermal conductivity. The dashed lines show the steady state temperature in the center of the inclusion as determined by the model of Goldenberg and Tranter, eqn. (2).

We observe that the infinite matrix results are in good agreement with the numerical results for the inclusion embedded within the film, see Figure 1, when k_F/k_S is of order unity. This would be the usual case were the film conductivity the same as the corresponding bulk value. When $k_F/k_S << 1$, the infinite matrix results overestimate the temperature as determined from the more accurate numerical model. When the inclusion is very conducting, $k_I/k_S >> 1$, the expressions of eqn. (2) overestimate the temperature in he inclusion. It is thus seen that the model of Goldenberg and Tranter [6] as used by Lange et al.[7,8], on which the scaling law of eqn. (1) is based, is not accurate when the film thermal conductivity is low compared to that of the substrate, and that the model of the inclusion in the infinite matrix does not always accurately describe the actual temperature in the inclusion even though the matrix is identified with the film material.

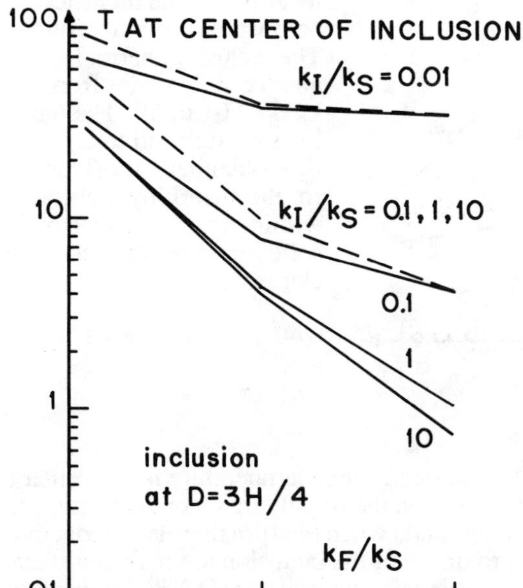

Figure 2. Steady state temperature in the center of the inclusion (measured in units of $QR^2/2k_S$) vs. the ratio of film to substrate thermal conductivity for several values of the inclusion thermal conductivity. The dashed lines correspond to eqn. (2) by Goldenberg and Tranter [6] for the absorbing inclusion within the infinite matrix. The solid lines are from the finite element solution.

Figure 3 shows the steady state temperature at the center of the film/substrate interface immediately below the inclusion. The result from the infinite matrix model is shown as calculated via eqn. (2) with the identification r=H-D>R. When compared to the numerical results, it is seen that this model overestimates the temperature at the center of the interface. The numerical result suggests that the temperature at the interface is not only low (as compared to the inclusion temperature), but it is also fairly independent of the film thermal conductivity. Thus, the experimentally measured low values for the thermal conductivity of thin films are not expected to affect significantly the temperature along the interface. In this context, we note the observation of Lowdermilk and Milam [2] that no correlation was found between the damage threshold and coating adhesion.

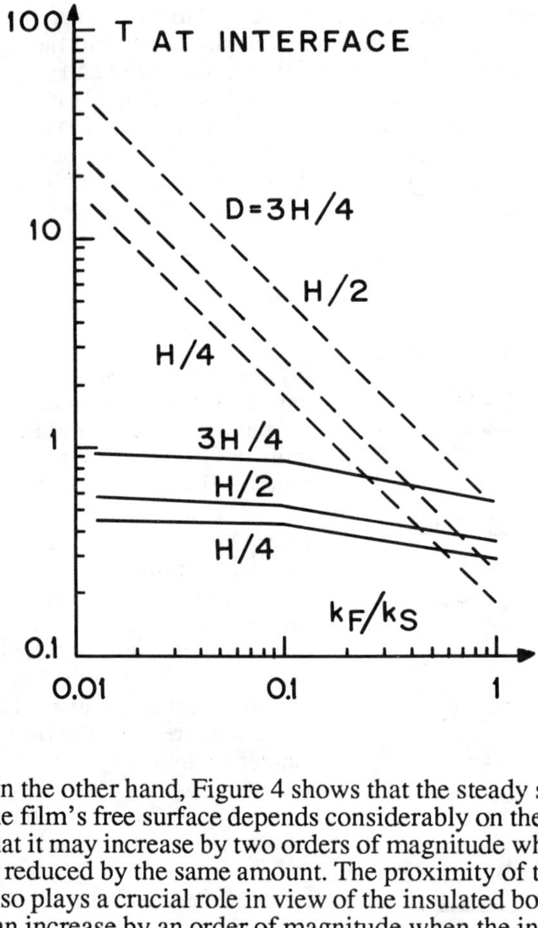

Figure 3. Steady state temperature in the center of the film/substrate interface r_1=0, z=-(H-D) (measured in units of $QR^2/2k_S$) vs. the ratio of film to substrate thermal conductivity for several locations of the inclusion below the film's free surface. The inclusion thermal conductivity varies from k_I/k_S=0.01 to 10. The dashed lines correspond to eqn. (2) by Goldenberg and Tranter [6] for the absorbing inclusion within the infinite matrix. The solid lines are from the finite element solution.

On the other hand, Figure 4 shows that the steady state temperature in the center of the film's free surface depends considerably on the film thermal conductivity, and that it may increase by two orders of magnitude when the film thermal conductivity is reduced by the same amount. The proximity of the inclusion to the free surface also plays a crucial role in view of the insulated boundary at z=D. The temperature can increase by an order of magnitude when the inclusion is moved from D=3H/4 to H/4. Comparison of Figure 2 to Figure 4 shows that another implication of low film thermal conductivity is that the temperature in the free surface becomes comparable to the temperature in the center of the inclusion when $k_F << k_S$. This is not the case when the film conductivity is comparable to that of the substrate. Thus, the present numerical model suggests that the film's free surface is another location where excessive temperatures may lead to laser damage, especially when the inclusion is situated near the free surface. We also note that the free surface temperature from the infinite matrix model, see eqn. (2), shows approximately the same dependence on k_F as the more accurate numerical solution.

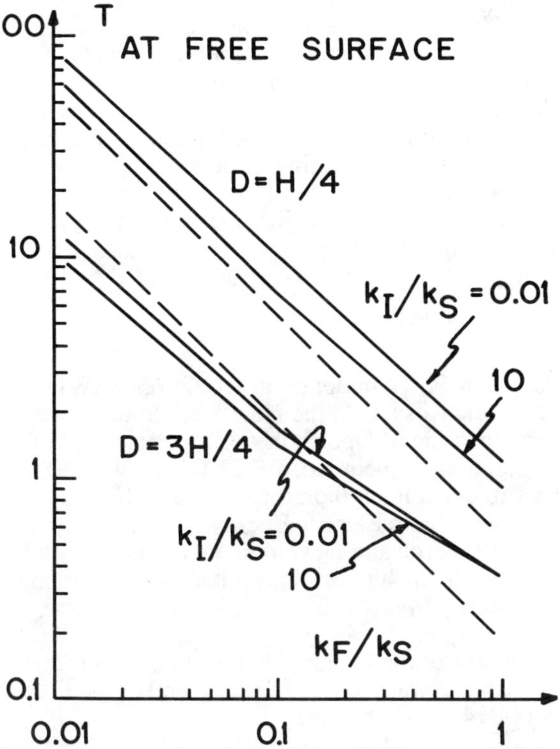

Figure 4. Steady state temperature in the center of the film's free surface at r_1=0, z=D (measured in units of $QR^2/2k_S$) vs. the ratio of film to substrate thermal conductivity for several locations of the inclusion below the film's free surface. The inclusion thermal conductivity varies from k_I/k_S=0.01 to 10. The dashed lines correspond to eqn. (2) by Goldenberg and Tranter [6] for the absorbing inclusion within the infinite matrix. The solid lines are from the finite element solution.

3. CONCLUSIONS

The very low values for the thermal conductivity of dielectric thin films imply that the predictions of the model of an absorbing inclusion embedded within a non-absorbing infinite matrix, and specifically eqn. (1) for the scaling of the critical energy density at damage with the film's conductivity, may not be accurate. Nevertheless, such a scaling establishes the importance of the film thermal conductivity in determining the laser damage resistance. Eqn. (2) or Figures 2-4 show that the temperature in the inclusion is very high when the film material has a low thermal conductivity compared to the substrate. When the thermal conductivity of the film is comparable to that of the substrate, then the model leading to eqn. (1) is accurate. In addition to the inclusion, the present work also identifies the importance of the free film surface as a possible site for damage when the film conductivity is low, and when the inclusion is located near the free surface of the film.

The more accurate model presented in this report accounts for the thickness of the film, for the proximity of the absorbing inclusion to the free surface and to the film/substrate interface, as well as for the different thermal conductivities of the film, inclusion, and substrate materials. The results presented refer to steady state temperatures which are known to be the largest. In that sense, the discrepancy discussed above between the model of the inclusion within the infinite matrix and the more accurate numerical model accounting for the presence of the film is largest under steady state conditions. Since laser damage is a time dependent phenomenon, it is clear that the variation of the temperature with time must be explicitly accounted in addition to the film thickness, inclusion size, and low film thermal conductivity in a manner analogous to the work of Lange et al. [7,8].

ACKNOWLEDGEMENTS

This work was supported by the U.S. Air Force under contract F2965088WUI64; Optical Coating Laboratory, Inc. of Santa Rosa, CA; the New York State Center for Advanced Optical Technology of the Institute of Optics; the U.S. Department of Energy Division of Inertial Fusion under agreement No. DE-FC03-85DP40200; and by the Laser Fusion Feasibility Project at the Laboratory for Laser Energetics which has the following sponsors: Empire State Electric Energy Research Corporation, New York State Energy Research and Development Authority, Ontario Hydro, and the University of Rochester. Such support does not imply endorsement of the content by any of the above parties.

The authors also acknowledge the support of the National Science Foundation under the Presidential Young Investigator Award MSM-8857096 and of the Office of Naval Research under Grant N-00014-87-K-0488.

REFERENCES

1. T.W. Walker, A.H. Guenther, and P.E. Nielsen, Pulsed laser-induced damage to thin-film optical coatings-Part I: Experimental, IEEE J. Quantum Electronics, QE-17, 2041 (1981).
2. W.H. Lowdermilk and D. Milam, Laser-induced surface and coating damage, IEEE J. Quantum Electronics, QE-17, 1888 (1981).
3. R.R. Austin, R. Michaud, A.H. Guenther, and J. Putman, Effects of structure, composition, and stress on the laser damage threshold of homogeneous and inhomogeneous single films and multilayers, Appl. Optics, 12, 665 (1973).

4. T.W. Walker, A.H. Guenther, and P.E. Nielsen, Pulsed laser-induced damage to thin-film optical coatings-Part II: Theory, IEEE J. Quantum Electronics, QE-17, 2053 (1981).
5. R.W. Hopper and D.R. Uhlmann, Mechanism of inclusion damage in laser glass, J. Appl. Phys., 41, 4023 (1970).
6. H. Goldenberg and C.J. Tranter, Heat flow in an infinite medium heated by a sphere, Brit. J. Appl. Phys., 3, 296 (1952).
7. M.R. Lange, J.K. McIver, and A.H. Guenther, The influence of the thermal and mechanical properties of optical materials in thin film form on the damage resistance to pulsed lasers, Thin Solid Films, 118, 49 (1984).
8. M.R. Lange, J.K. McIver, and A.H. Guenther, Pulsed laser damage in thin film coatings: fluorides and oxides, Thin Solid Films, 125, 143 (1985).
9. A.H. Guenther and J.K. McIver, The role of thermal conductivity in the pulsed laser damage sensitivity of optical thin films, Thin Solid Films, 163, 203 (1988).
10. J.C. Lambropoulos, M.R. Jolly, C.A. Amsden, M. Sinicropi, D. Diakomihalis, and S.D. Jacobs, Thermal conductivity of dielectric thin films, J. Appl. Phys., 66, 4230 (1989).
11. R.W. Powell, Thermal conductivity determinations by thermal comparator methods, in: Thermal Conductivity, vol. 2, p. 276 (ed. R.P. Tye), Academic Press, New York (1969).

Section IV. Device Applications

PHOTOREFRACTION: SOME APPLICATIONS

Edward J. Sharp, William W. Clark,III,
Mary J. Miller, and Gary L. Wood
CECOM Center for Night Vision & Electro-Optics
Fort Belvoir, Virginia 22060-5677

Gregory J. Salamo and Brian D. Monson
University of Arkansas
Fayetteville, Arkansas 72701

Ratnakar R. Neurgaonkar
Rockwell International Science Center
Thousand Oaks, California 91360

ABSTRACT

In this paper, the photorefractive nonlinearity is explored in tungsten bronze family ferroelectric crystals. Applications which take advantage of both the amplified and deamplified beams in the photorefractive process will be addressed. In particular, the discussion will include details of the photorefractive optical nonlinearity and its use in **optical limiting**, **beam clean-up** via phase conjugation, the exchange of information between optical beams which are mutually incoherent through **double phase conjugation**, and **multi-colored photorefractive effects**.

INTRODUCTION

The term "photorefraction" denotes a refractive index change which is induced by light. Although there are many mechanisms which can lead to the index change, "photorefraction" is usually reserved for index changes caused by light-induced charge migration in electro-optic materials. There are a number of factors which set photorefractive index changes apart from other light-induced index changes. Briefly, above some saturation intensity, the magnitude of the steady-state photorefractive index change is independent of the total intensity of the incident light; only its time for formation depends on the light intensity. The maximum photorefractive index change does not necessarily occur where the beam intensities are greatest or, in other words, the index change is spatially non-local. By contrast, the index changes giving rise to optical "self-actions" such as stimulated scattering, self-focusing and self-defocusing scale with increasing beam power or intensity. These changes are spatially local (largest index changes occurring in regions of highest intensity), and can occur on picosecond time scales. However, the time of formation is, in general, independent of beam intensity. Materials which exhibit photorefractive behavior are important for a number of applications. In the sections that follow we will provide a brief review of the photorefractive effect, describe techniques to characterize photorefractors, discuss the material parameters considered important for applications, and present recent research results in several application areas.

THE PHOTOREFRACTIVE EFFECT

The microscopic details of the photorefractive mechanism is normally described using a band transport model[1-4] which assumes the existence of a pool of charges residing in low-lying traps. For example, in the presence of nonuniform or spatially periodic illumination (Fig. 1a), the charges (either electrons or holes) are excited from defect sites or impurity centers leaving behind ionized donor sites. The photo-excited charges migrate by diffusion (Fig. 1b) out of the illuminated regions and are eventually retrapped in the dark regions of the crystal. A static charge distribution (Fig. 1c) is formed in the crystal due to the spatial separation of the ionized donor sites in the illuminated regions and the retrapped charges in the dark regions. This space-charge distribution is balanced by a strong space-charge field according to Poisson's equation. The space-charge-field distribution (Fig. 1d) is shifted spatially ($\pi/2$ degrees) relative to the incident intensity distribution. This strong electrostatic field (~10^5 V/m) then produces a change in the refractive indices ($\delta n \approx 10^{-3}$) through the linear electro-optic effect (Fig. 1e). If the incident intensity pattern is periodic (i.e., the intensity pattern is produced by interfering two coherent laser beams) then the resulting index change will also be periodic, resulting in the formation of an "index grating." The $\pi/2$ phase shift between the incident intensity pattern and the index grating causes the two interfering or "writing" beams to couple as they propagate through the crystal. In this two-wave mixing process one of the beams will experience gain and the other loss. Feinberg, et al., demonstrated that one beam can deplete essentially all of the power of the other beam after propagating through only a few millimeters of a $BaTiO_3$ crystal.[5]

As seen from this description of the photorefractive effect, photorefractive materials can best be characterized by their ability to produce an index change via charge transport. A given material, therefore, can be characterized by the **magnitude** of the index change produced, the **speed** with which that index change occurs, and the crystal **dark current**, which determines how long the index change will be preserved. These three characteristics can be determined experimentally for a given material. Likewise, a material can be designed or altered to have specific performance in these areas.

One of the more common ways to characterize a photorefractive material is by two-beam coupling experiments. The apparatus used for these measurements is shown in Fig. 2a. The incident beam intensity is chosen such that $I_1 \ll I_2$. In this limit of small modulation m, I_1 experiences gain via energy exchange according to;

$$I_{1c}/I_1 = \exp(\Gamma l), \qquad [1]$$

where l is the interaction length, I_1 is the incident weak beam, I_{1c} is the intensity of beam 1 with steady-state coupling, and Γ the two-beam coupling gain coefficient

$$\Gamma = \frac{2\pi r_{eff} F R K_B T K_g}{(\lambda q n)\cos(\Theta)[1 + (K_g/K_o)^2]}. \qquad [2]$$

Here $K_o^2 = q^2 N_{eff}/K_B T \varepsilon\varepsilon_o$, $K_g = 2(\omega/c)\sin\Theta$ is the grating wave vector, $K_B T$ is the Boltzmann thermal energy, $N_{eff} = N_A(1 - N_A/N_D)$ is the effective density of charge carriers, N_D is the number of donor sites, N_A is the number of trap sites under dark conditions and $N_D > N_A$, R is the electron-hole competition number (-1 to +1), F is the fractional poling factor for ferroelectric crystals (0 to 1), and r_{eff} is the effective electro-optic coefficient which depends on the crystal symmetry as well as the orientation. For crystals with 4mm symmetry and extraordinary light,

$$r_{eff} = n_e^4 r_{33}\cos^2\Theta - n_o^4 r_{13}\sin^2\Theta, \qquad [3]$$

while for ordinary light,

$$r_{eff} = n_o^4 r_{13}. \qquad [4]$$

Figure 2b shows the normalized experimental points of the ratio I_{1c}/I_1 (ordinary light) measured as a function of external angle Θ for ferroelectric Ce:SBN crystals plotted as K_g/Γ versus K_g^2.[6] The slope-intercept form of Eq. [2] can be written as;

$$\frac{K_g}{\Gamma(n\omega/c)} = \frac{\varepsilon\varepsilon_o n \cos\Theta}{q N_{eff} r_{eff} F R}\,\frac{K_g^2}{(n\omega/c)^2} + \frac{n q \cos\Theta}{r_{eff} F R K_B T (n\omega/c)^2} \qquad [5]$$

N_{eff} is determined from the slope of the curve, while the intercept yields the value of $r_{eff}R$. These measurements therefore yield the magnitude of N_{eff} and R, both of which are instrumental in determining the value of the space charge field and the magnitude of the index change. Here it is assumed that the electro-optic coefficients are known or are determined by a separate measurement. The data shown in Fig. 2b are for two different samples of SBN:60 at the 632.8 nm laser wavelength of HeNe.

The experimental set up of Fig. 2a is also used to measure the time response, dark conductivity, mobility-recombination rate product, and also N_{eff}. By monitoring the energy coupled as a function of time the photorefractive response time can be determined (grating formation). These measurement show that the response time varies inversly with the intensity. The response time can also be measured by monitoring the grating erasure time produced by a uniform erasing beam. If the writing beams are adjusted such that the grating spacing is kept much greater than the diffusion length, a plot of the erasure time versus incident intensity yields a value for the dark conductivity (intercept) and the product of the mobility and recombination time (slope). Finally, the dark decay rate of a photorefractive grating has a strong temperature dependence. A plot of the dark decay rate as a function of the grating spatial frequency squared (K_g^2), at a constant temperature, will give a value of the dark conductivity from the intercept and a value of N_{eff} from the slope.

Other important material considerations which affect the photorefractive process are the magnitude and sign of the electro-optic coefficients, the dielectric constants, the crystal symmetry, absorption, optical quality, and crystal size.

A number of experiments have been conducted recently which point out the versatility of the tungsten bronze class of ferroelectric crystals for photorefractive applications. Large doped crystals of the tungsten bronze family such as SBN, BSKNN, and SCNN, have been grown with high optical quality.[7-9] These crystals possess high gain, an intensity dependent time response (presently at 50 to 100 msec for $1W/cm^2$), and low dark conductivity. These qualities make them attractive for many applications in the areas of phase conjugation, optical processing and beam control devices. The high sensitivity of photorefractive materials has made it possible to demonstrate all of these applications using very weak optical beams (milliwatt). To date, this type of performance has only been possible through the use of the "photorefractive nonlinearity."

The above qualities of tungsten bronze crystals depend on the strongly on the ability to grow large high optical quality single crystals. This can be an expensive and time consuming undertaking. Photorefractive ceramics with high gain, fast time response, and good transparency would be extremely useful for many applications and attractive for their potential low cost compared to single crystal photorefractors. To achieve these qualities photorefractive ceramics need: (a) many domain directions (for easy poling), i.e., SBN has 180° or two domains while $BaTiO_3$ has 90° and 180° or four + two domains, (b) low sintering and pressure (for easy poling), and (c) high transparancy so the material can be as thick as possible for long interaction lengths.

APPLICATIONS

In materials with high gain such as $BaTiO_3$, SBN, and BSKNN it is easy to observe beam fanning (asymmetric beam scattering)[10] and degenerate four-wave mixing.[11] There are a number of four-wave mixing configurations possible with photorefractive materials but perhaps the most surprising is self-pumped phase conjugation via internal reflection [12] and double-pumped phase conjugation.[13] Beam fanning, self-pumped phase conjugation, and double phase conjugation have also been demonstrated for mutually incoherent laser beams (multicolor effects).[14,15] Applications arising from each of these observed phenomena will be reviewed in this section.

Optical Limiting

As mentioned earlier, the throughput transmission of beam #2 (Fig. 2a) is decreased by the two beam coupling process. This can serve as the basis for an optical limiter. In this configuration a small amount of the incoming radiation is split off and made to cross the main beam in the crystal. For incoherent light no coupling will occur, but coherent light will couple with the weak beam and be deflected in that direction. Thus the coherent radiation is limited in the direction of the strong incoherent beam, while the incoherent beam undergoes no such limiting action.

A more simple optical limiter can be devised which takes advantage of the beam fanning[10] process. In beam fanning a single beam of coherent light is scattered asymmetrically as it passes through a high gain crystal (Fig. 3a). In this process a small fraction of the incident beam (beam 1) is scattered in all directions by crystal imperfections (beams 2). These beams propagate across the incident beam as depicted. For one direction (relative to the c-axis) the scattered beams experience gain via two-beam coupling and are amplified at the expense of the incident beam. In the other direction the scattered light couples back into the main beam. This results in light being preferentially scattered to one side of the crystal which significantly reduces the intensity of the incident beam and therefore acts as an **optical limiter.**[16,17]

The important parameters for optical limiting are 1) the experimental decay times of the beams transmitted through a photorefractive crystal and 2) the steady state transmission, i.e., the optical transmission as time goes to infinity, T_{inf}. The transmitted intensity through a beam fanning limiter as depicted in Fig. 3a can be approximated as;

$$I_t = I_o(1 - T_{inf})\exp(-t/T_{NL}) + T_{inf}I_o, \qquad [6]$$

where I_o is the input intensity and T_{NL} is the photorefractive response time of the crystal. The exponential decay of I_o in Eq. [6] is an approximation which roughly resembles the observed decay. Integration of Eq. [6] from time t = 0 to t_p (i.e., over the width of the pulse) gives the transmitted energy density, R_t,

$$R_t = A(1 - T_{inf})[1 - \exp(-t_p/T_{NL})] + R_{in}T_{inf} \quad [7a]$$

or,

$$R_t = A(1 - T_{inf})[1 - \exp(-R_{in}/A)] + R_{in}T_{inf} \quad [7b]$$

where $R_{in} = I_o t_p$ is the input energy density and $T_{NL} = A/I_o^x$ with $x \sim 1$. A is a wavelength-dependent material constant with units of energy density.[18] As can be seen, photorefractive beam fanning is important for $t_p \gtrsim T_{NL}$ or for $R_{in} \gtrsim A$. The value of A determines the energy density where limiting occurs and must be below the value of the detector damage level for effective limiting. The value of A depends on the crystal composition, the dopant species and concentration, as well as the presence of external electric fields. An A value of 0.1 J/cm^2 along with and a T_{inf} value of .001 has been measured for beam fanning in a Cerium doped strontium barium niobate crystal when used to limit a helium cadmium laser at 442 nm.[19]

Beam Clean-up

A convenient geometry used in many phase conjugation applications is the self-pumped phase conjugate mirror (SPPCM) configuration where the phase conjugate mirror is simply a photorefractive single crystal.[12] In this configuration a portion of the beam fan is shown schematically as emanating from two locations along the incident beam in Fig. 3b. These beams strike the corner of the crystal and are totally internally reflected. The reflected radiation interferes with the incident radiation such that four-wave mixing occurs. In this figure the reflected waves (3 and 3') which make up the corner loop are shown being scattered as phase-conjugate waves (4 and 4') from gratings written by the scattered waves (2 and 2') and the incident waves (1 and 1'). This feedback arrangement strengthens the index gratings at two locations deep within the crystal.[20] Figure 3c shows a photograph of the loop formed in an SBN:60 crystal while it is self pumping.

The shape and position of the loop within the crystal can vary widely from crystal to crystal. The void (lack of light) between the beams forming the loop can be wide (~ a millimeter) to barely discernible depending upon the angle and position of the incident beam and the condition of the crystal faces adjacent to the corner containing the loop.

This SPPCM is self-starting (arises from beam fanning) and self-aligning (produces a phase conjugate replica for a wide range of angles of incidence). It can operate using only milliwatt laser beam powers. The features of self-alignment and low operating powers are distinct advantages for beam clean-up applications since these eliminate the alignment problems associated with four-wave mixing and the high intensities (short pulse) requirements imposed by stimulated Brillioun scattering. In fact, because of the high gain ($\Gamma l > 3$) associated with the photorefractive effect in materials like $BaTiO_3$ and tungsten bronze crystals, demonstrations of SPPCM's with milliwatt laser beams has been thus far restricted to photorefractors.

Using the experimental set-up depicted in Fig. 4, we verified that the backscattered wave was a phase-conjugate of the input wave. This was accomplished by using the distortion correction **(beam clean-up)** property of the conjugate wave.[21] Figure 4a shows the input scene located in the object plane (transparency) which is imaged by lens L_1 into the crystal. The reflected beam is then re-imaged by L_1 and L_2 onto the image plane (screen) and displayed in Fig. 4b. As can be seen, the image is a faithful reproduction of the object. In fact, the image is the phase conjugate of the input scene. This is clearly demonstrated by adding a wavefront distortion to the input beam at D_1. The phase-conjugate reflected beam examined at the screen does not retrace its path through the aberrator, and hence, cannot become undistorted as shown in Fig. 4c. However, when the wavefront distorter is placed at D_2 the phase-conjugated reflected beam does retrace is path through the distorter and becomes undistorted as seen in Fig. 4d. The reconstruction of the object after a double pass through a distorter verifies that the image seen in Fig. 4d is indeed the phase conjugate.

Double Phase Conjugation

A recent development in phase conjugation techniques has been the double phase conjugate mirror (DPCM)[13,22] where the photorefractive material acts as a conjugate mirror for two incoming waves simultaneously as shown in Fig. 5. When the output of two independent lasers with incident optical electromagnetic fields E_1 and E_2 are directed into a photorefractive medium, a phase conjugate replica of each beam appears under appropriate conditions. Both beams are required for either conjugate (E_1^* or E_2^*) to exist and the energy for the conjugate of one beam is supplied by the other beam. For example, if E_1 is blocked, the conjugate E_2^* disappears instantly, while the conjugate E_1^* disappears slowly with the characteristic erasure time of the crystal. A similar behavior is observed if E_2 is blocked. In this conjugator, beam 1 writes the gratings responsible for the production of its conjugate while beam 2 reads these gratings and supplies the energy for the conjugate wave.

DCPMs require no external pumping beams and the two incident beams need not be coherent with one another, although the sources must have nearly the same wavelength. They also have the following additional properties;

1. Conjugation of two beams occurs simultaneously;
2. The two conjugating beams may be mutually incoherent;
3. No cross talk is observed between conjugating beams;
4. Digital information can be immediately transferred between beams;
5. Images can be conjugated with excellent fidelity;
6. Subtraction and addition of images can be performed.

These properties make the double phase conjugators attractive for applications in free space communications, optical interconnections, beam clean-up, and optical image processing. The experiment shown in Fig. 5 was used to demonstrate image addition/subtraction and show that the bridge conjugator can produce conjugate images with high fidelity. The conjugate of the object slide, T, seen at screen S_1 is shown in Fig. 5a. The image is an exact replica of the object slide. Figure 5b shows the output of BS_2 when projected onto S_2. As can be seen, Fig. 5b is an exact negative of the object slide. This occurs due to destructive interference between the conjugate signals E_1^* and E_2^*. E_2^* has a near constant intensity while E_1^* is the conjugate of the input image. As a result of the 180° phase shift introduced by BS_2, these two beams destructively interfere when imaged together on S_2. As expected this result persisted even when a constant phase shift was placed in the path of E_2 or when the optical path of beam E_2 was physically changed. This method of image subtraction is much less involved than others previously reported in photorefractive materials.[25] The image produced at S_1 is the sum of E_1^* and E_2^* due to constructive interference at BS_1.

Although there are many different configurations for DPCMs,[13,22] there are significant benefits to be derived from specific geometric arrangements. For example, both the time response and magnitude of the conjugate signal will vary significantly depending on which geometry is employed for a particular crystal. Recently we reported[23,24] on a new method for double phase conjugation which is particularly suited to the tungsten-bronze crystal strontium barium niobate (SBN), although the method has also been applied to $BaTiO_3$ and BSKNN. This new arrangement is highly insensitive to the alignment of the incident beams, quick to form a conjugate image with high reflectivity and fidelity, and free from instabilities due to frequency shifts or competition with self-pumping. The geometry for this new DPCM is depicted in Fig. 5 where the two input beams have "fanned" into one another to form the coupling region observed in the crystal after the conjugate waves appear (represented by the dashed line within the crystal). Since the two beams bridge to overlap in the crystal without reflecting off a crystal face, this arrangement is called the "bridge conjugator."

Multicolor Photorefractive Effects

Other interesting effects have been observed by directing multicolored light from a coherent source into a photorefractive crystal. For instance, a rainbow pattern was formed in the beam fan when the crystal was inserted into a multiline argon-ion laser beam. The origin of this effect is Bragg scattering off selectively enhanced photorefractive gratings.[14] When the multicolored ring was reflected back into the crystal the phase conjugate of all eight laser lines was produced in a time very near to the beam fanning response time.[15] It was also possible to produce multi-lined self-pumped phase conjugation via internal reflection at several different wavelengths by spatially separating the argon-ion laser lines with a dispersing element. These discoveries have made it possible to demonstrate real-time color holography.[27]

Figure 6 shows the experimental setups used to observe normal beam fanning as well as the rainbow pattern. Projected sketches of the normal beam fan and rainbow pattern are depicted in Figs. 6a and 6b respectively. As shown in these figures, markedly different effects were obtained by rotating the crystal 90°. The output of an argon-ion laser, oscillating at eight discrete lines simultaneously, is directed onto an SBN or BSKNN crystal. The unfocused incident beam makes an angle θ with the normal to the entrance crystal face and is linearly polarized. The optic axis of the crystal is in the plane of incidence and with a positive direction indicated by the vector c. A broad fan (normal beam fan) of linearly polarized light appears on the negative c-axis side of the transmitted beam. This fan appears independent of whether θ is positive or negative as defined by the c direction. For negative θ values, however, an intense multi-colored ring or rainbow pattern also appears as shown in Fig. 6b. The ring is observed for either ordinary or extraordinary polarized light with the polarization of the ring having the same polarization as the incident light. However, the observed pattern was weaker for ordinary polarized light.

Examination of the rainbow pattern using a spectrometer indicates that all eight lines are present, although only four or five of them are discernible to the eye. In fact, reflection of the multi-colored ring back into the crystal using a curved mirror produced the phase conjugate of all eight lines in a time very near to the crystal beam-fanning response time. In this sense, the crystal acts as a broadband "self-pumping" device.[26] Multicolor phase conjugate signals were also observed when a pair of dispersing prisms were used to separate the beams so they could self pump independently in the crystal. The details of these observations is the subject matter of another paper in these proceedings.[27]

CONCLUSIONS

The important features which set the photorefractive nonlinearity apart from other optical nonlinearities have been presented and the more relevant photorefractive material parameters have been identified. Methods to experimentally determine the gain, time response, and dark conductivity are described. Selected applications which demonstrate the effective use of photorefractive beam fanning, self-pumped phase conjugation, double phase conjugation, and multicolor phase conjugation are given including their use in real-time color holography, free space comunications, optical image-processing, optical interconections, laser hardening, image amplification, and aberration correction.

REFERENCES

1 N. V. Kukhtarev, V. B. Markov, S. G. Odulov, M. S. Soskin, and V. L. Vinetskii, "Holographic Storage in Electrooptic Crystals. I. Steady State," Ferroelectrics **22**:949-960 (1979).

2 G. C. Valley and M. B. Klein, "Optimal Properties of Photorefractive Materials for Optical Data Processing," Opt. Eng. **22**:704-711 (1983).

3 F. P. Strohkendl, J. M. C. Jonathan, and R. W. Hellwarth, "Hole-Electron Competition in Photorefractive Gratings," Opt. Lett. **11**:312-314 (1986).

4 P. Gunter, "Holography, Coherent Light Amplification and Optical Phase Conjugation with Photorefractive Materials," Phys. Rep. **93**:199-299 (1982).

5 J. Feinberg, D. Heiman, A. R. Tanguay, Jr., and R. W. Hellwarth, "Photorefractive Effects and Light Induced Charge Migration in Barium Titanate," J. Appl. Phys. **51**:1297-1305 (1980).

6 G. L. Wood, W. W. Clark, III, M. J. Miller, E. J. Sharp, G. J. Salamo, and R. R. Neurgaonkar, "Broadband Photorefractive Properties and Self-Pumped Phase Conjugation in Ce-SBN:60," IEEE J. Quantum Electron. **QE-23**:2126-1235 (1987).

7 R. R. Neurgaonkar, W. K. Cory, J. R. Oliver, E. J. Sharp, M. J. Miller, W. W. Clark, III, G. L. Wood, and G. J. Salamo, "Cr^{3+}-$Sr_{0.6}Ba_{0.4}Nb_2O_6$ Single Crystals for Photorefractive Applications," Mat. Res. Bull. **24**:589-594 (1989).

8 R. R. Neurgaonkar, W. K. Cory, J. R. Oliver, W. W. Clark, III, G. L. Wood, M. J. Miller, and E. J. Sharp, "Growth and Ferroelectric Properties of Tungsten Bronze $Ba_{2-x}Sr_xK_{1-y}Na_yNb_5O_{15}$ (BSKNN) Single Crystals," J. Cryst. Growth **84**:629-637 (1987).

9 R. R. Neurgaonkar, W. K. Cory, J. R. Oliver, E. J. Sharp, G. L. Wood, M. J. Miller, W. W. Clark, III, and G. J. Salamo, "Growth and Ferroelectric Properties of Tungsten Bronze $Sr_{2-x}Ca_xNaNb_5O_{15}$ Single Crystals," Mat. Res. Bull. **23**:1459-1467 (1988).

10 J. Feinberg, "Asymmetric Self-Defocusing of an Optical Beam from the Photorefractive Effect", J. Opt. Soc. Am. **72**:46-51 (1982).

11 J. Feinberg and R. W. Hellwarth, "Phase Conjugating Mirror with Continuous Wave Gain," Opt. Lett. **5**:519-521 (1980); erratum **6**:257 (1981).

12 J. Feinberg, "Self-Pumped, Continuous Wave Phase Conjugation Using Internal Reflection," Opt. Lett. **7**:486-488 (1982).

13 S. Weiss, S. Sternklar, and B. Fischer, "Double Phase-Conjugate Mirror: Analysis, Demonstration, and Application," Opt. Lett. **12**:114-116 (1987).

14 G. J. Salamo, M. J. Miller, W. W. Clark, III, G. L. Wood, E. J. Sharp, and R. R. Neurgaonkar, "Multicolored Effects in Photorefractive Crystals," in Technical Digest of the Conference on Lasers and Electro-Optics (Optical Society of America, Washington D.C., 1988), p. 60-61.

15 G. J. Salamo, M. J. Miller, W. W. Clark, III, G. L. Wood, E. J. Sharp, and R. R. Neurgaonkar, "Photorefractive Rainbows," Appl. Opt. **27**:4356-4358 (1988).

16 M. Cronin-Golomb and A. Yariv, "Optical Limiters Using Photorefractive Nonlinearities," J. Appl. Phys. **57**:4906-4910 (1985).

17 G. L. Wood, W. W. Clark, III, M. J. Miller, G. J. Salamo, and E. J. Sharp, "Evaluation of Passive Optical Limiters and Switches," SPIE Proc. **1105**:154-180 (1989).

18 M. J. Miller, E. J. Sharp, G. L. Wood, W. W. Clark, III, G. J. Salamo, and R. R. Neurgaonkar, "Time Response of a Cerium-Doped $Sr_{0.75}Ba_{0.25}Nb_2O_6$ Self-Pumped Phase Conjugate Mirror," Opt. Lett. **12**:340-342 (1987).

19 G. J. Salamo, M. J. Miller, W. W. Clark, III, G. L. Wood, and E. J. Sharp, "Strontium Barium Niobate as a Self-Pumped Phase Conjugator," Opt. Commun. **59**:417-422 (1986).

20 K. R. MacDonald and J. Feinberg, "Theory of a Self-Pumped Phase Conjugator with Two Interaction Regions," J. Opt. Soc. Am. **73**:548-553 (1983).

21 E. J. Sharp, M. J. Miller, G. J. Salamo, W. W. Clark, III, G. L. Wood, and R. R. Neurgaonkar, "Tungsten Bronze Crystals for Photorefractive Self-Pumped Phase Conjugation," Ferroelectrics **77**:335-346 (1988).

22 P. Yeh, T.Y. Chang, and M.D. Ewbank, "Model for Mutually Pumped Phase Conjugation," J. Opt. Soc. Am. B **5**:1743-1749 (1988).

23 G. J. Salamo, M. J. Miller, W. W. Clark, III, G. L. Wood, E. J. Sharp, and R. R. Neurgaonkar, "Double Phase Conjugation in Strontium Barium Niobate," in Technical Digest 1988 Annual Meeting of American Optical Society, (Optical Society of America, Washington D.C., 1988) p. 168.

24 E. J. Sharp, W. W. Clark, III, M. J. Miller, G. L. Wood, B. D. Monson, G. J. Salamo, and R. R. Neurgaonkar, "Double Phase Conjugation in Tungsten Bronze Crystals," Appl. Opt. **29**:743-749 (1990).

25 P. Yeh, T. Y. Chang, and P. H. Beckwith, "Real-Time Optical Image Subtraction Using Dynamic Holographic Interference in Photorefractive Media," Opt. Lett. **13**:586-588 (1988).

26 M. Cronin-Golomb, S. Kwong, and A. Yariv, "Multicolor Passive (Self-Pumped) Phase Conjugation," Appl. Phys. Lett. **44**:727-729 (1984).

27 G. J. Salamo, B. D. Monson, E. J. Sharp, W. W. Clark, III, M. J. Miller, G. L. Wood, and R. R. Neurgaonkar, "Color Imaging Using Tungsten Bronze Ferroelectric Crystals", These proceedings.

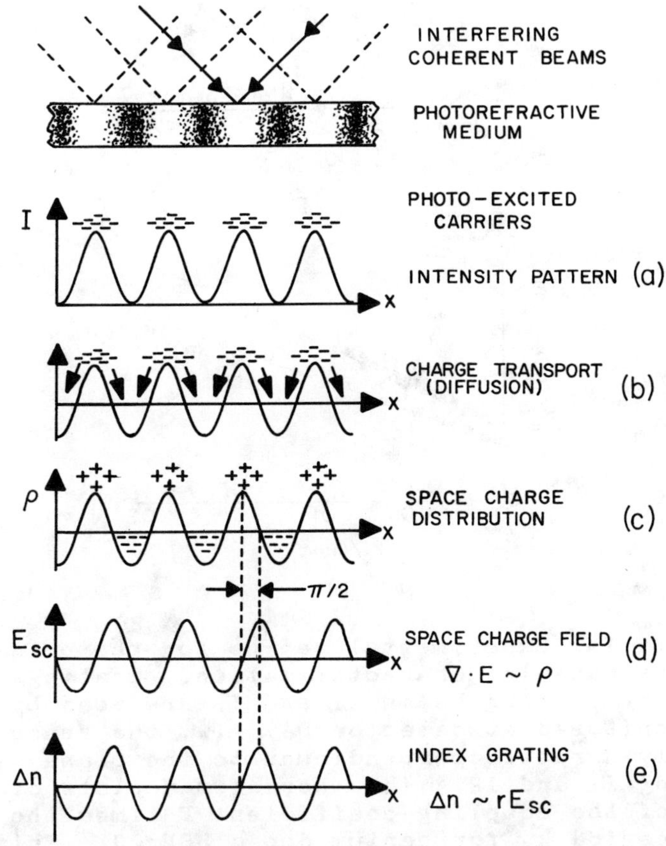

Figure 1. The formation of photorefractive index gratings. (a) Periodic light intensity pattern and photo-excited charge carriers. (b) Charge transport by diffusion. (c) Static charge distribution. (d) $\pi/2$ phase-shifted space charge field distribution. (e) Photorefractive index gratings.

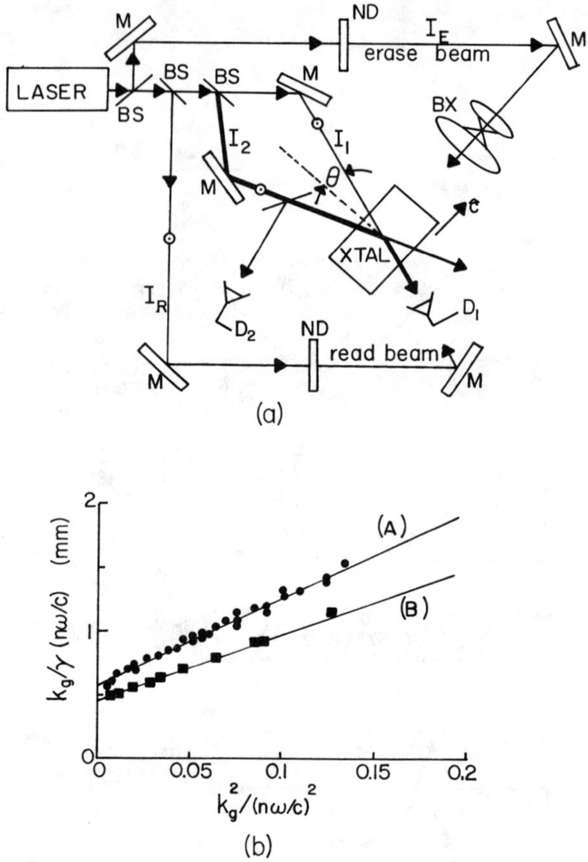

Figure 2. (a) Experimental set-up for photorefractive material characterization. Gratings produced by writing beams I_1 and I_2 are read by beam I_R and monitored at detector D_2. I_E, the erase beam, is usually directed perpendicuar to the plane containing I_1 and I_2. (b) Normalized plots of the inverse of the coupling coefficient Γ times the grating period k_g for cerium doped SBN:60. This data was obtained by monitoring the weak beam (I_1) gain at detector D_1 as a function of beam crossing angle. The slope-intercept form of the two-beam coupling gain coefficient Γ is used to extract N_{eff} and R. Curve (A) yields R = 0.45 and N_{eff} = 6.6 X 10^{16} cm^{-3}. Curve (B) yields R = 0.62 and N_{eff} = 7.0 X 10^{16} cm^{-3}. BS (beam splitter), BX (beam expander, ND (neutral density filters), and M (mirrors).

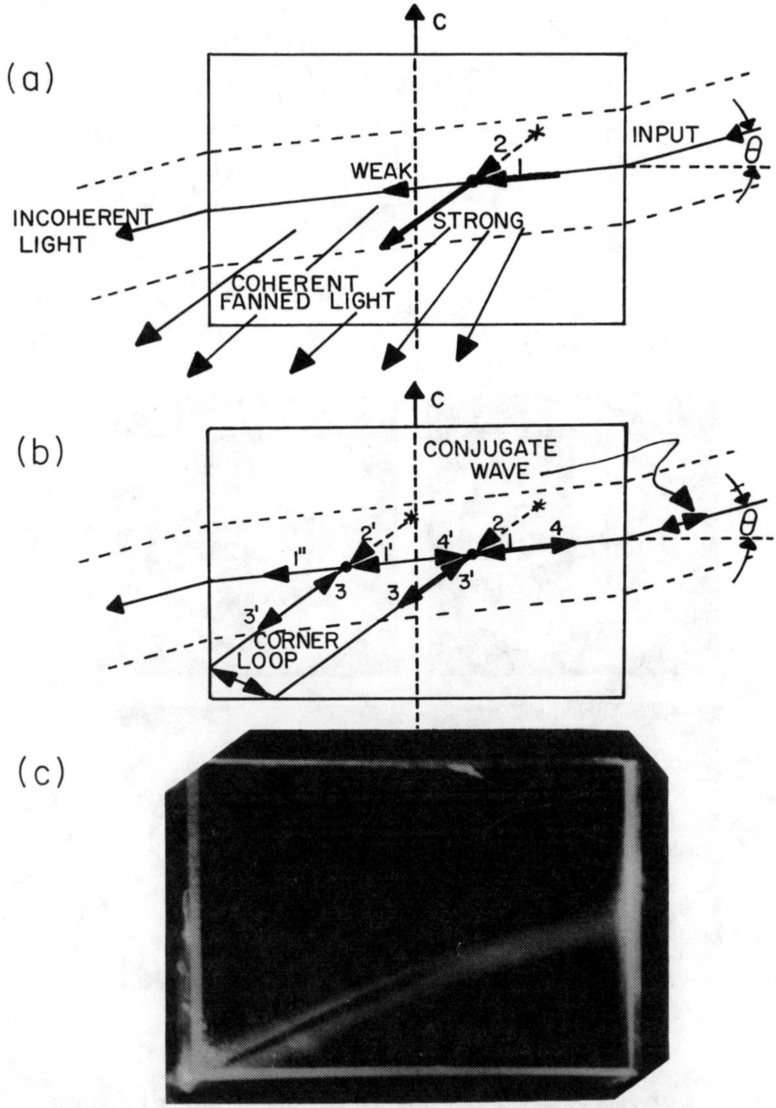

Figure 3. (a) Schematic of an optical limiter based on beam fanning produced by two-beam coupling. (b) Retro-reflected beam-fan light forming a corner loop and the resulting four-wave mixing leading to the phase-conjugate waves 4 and 4'. (c) Photograph of an SBN:60 crystal while self pumping.

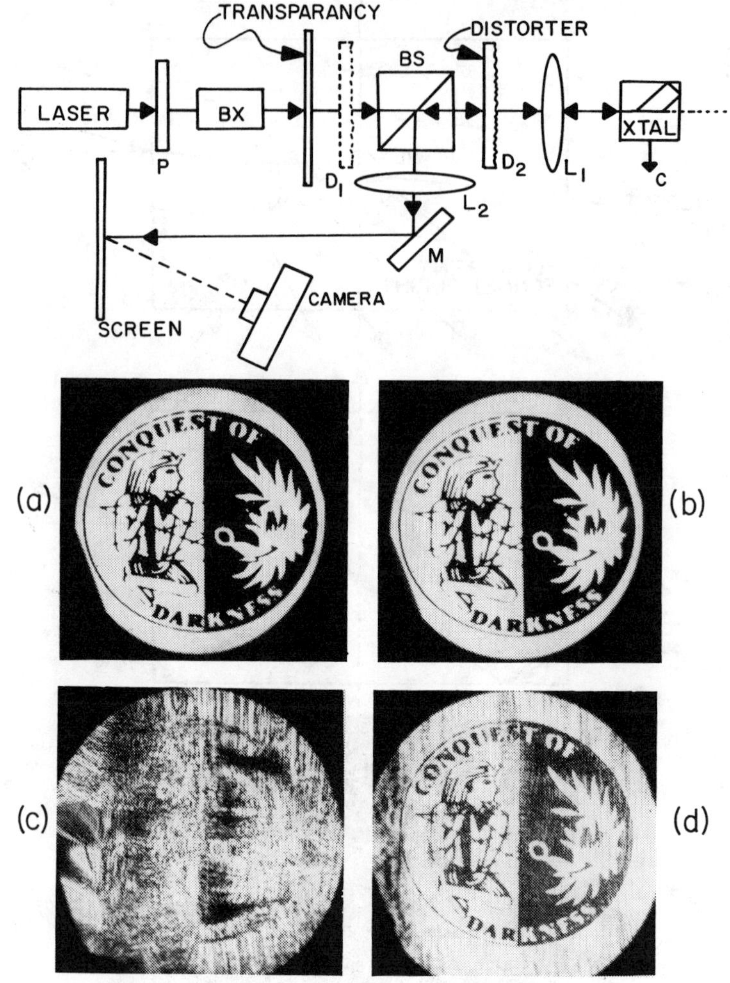

Figure 4. Schematic of apparatus used to produce phase conjugate replicas of input images and to test the aberration correction ability of self-pumped phase conjugate mirrors where BX is a beam expander and P a polarization rotator. (a) input image, (b) phase-conjugate replica of the input image, (c) phase-conjugate replica of the aberrated input image, and (d) phase-conjugate replica of the input image after aberration correction.

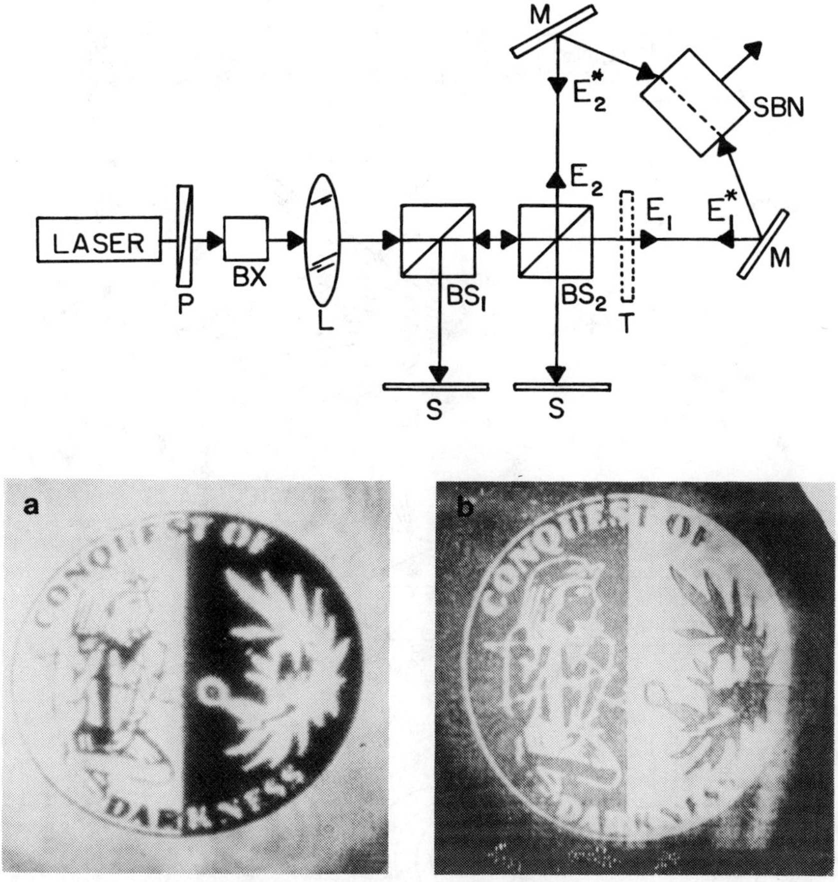

Figure 5. Experimental apparatus used to observe addition and subtraction of images with the "bridge conjugator" double phase conjugate mirror. P (polarizer), BX (beam expander), BS (beam splitter), and T (transparancy). (a) Conjugate replica of the object slide seen at screen S_1. (b) Conjugate replica of the negative of the object slide seen at S_2.

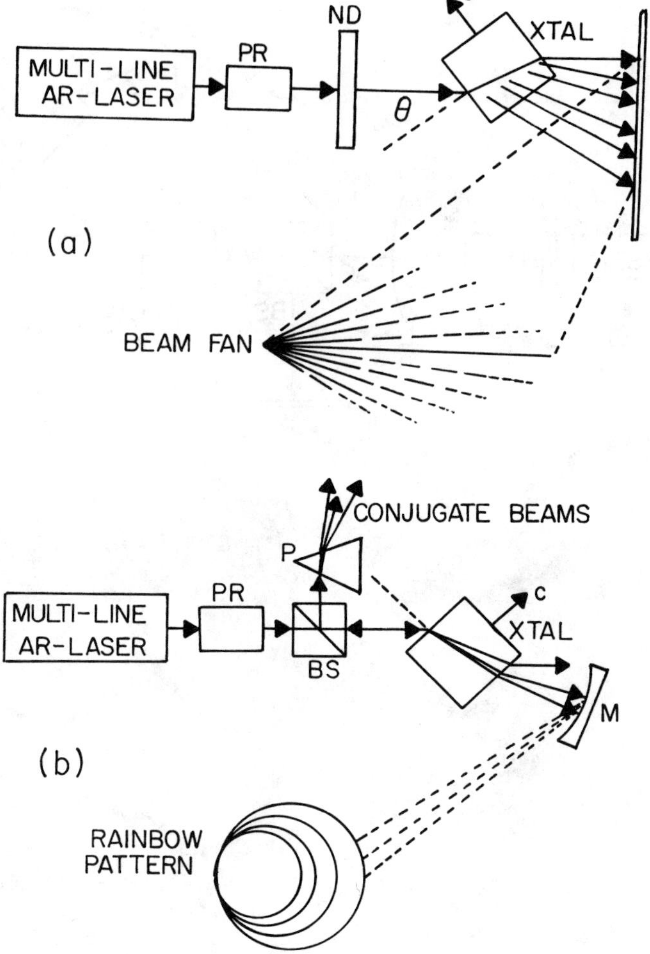

Figure 6. Experimental set-ups used to observe normal beam fanning and enhanced beam fanning or multicolor isotropic conical diffraction rings. In (b) a prism or grating is used to separate the conjugate waves obtained by focusing the rainbow pattern back into the crystal.

COLOR IMAGING USING TUNGSTEN BRONZE FERROELECTRIC CRYSTALS

Gregory J. Salamo and Brian D. Monson
University of Arkansas
Physics Department
Fayetteville, AR 72701

Edward J. Sharp, William W. Clark, III, Mary J. Miller and Gary L. Wood
CECOM Center for Night Vision and Electro-Optics
Fort Belvoir, Virginia 22060-5677

R.R. Neurgaonkar
Rockwell Science Center
Thousand Oaks, CA 91360

ABSTRACT

Color phase-conjugate imaging is demonstrated using a multi-colored laser beam and several tungsten bronze ferroelectric photorefractive crystals. Speed of response, size of the image, clarity of the image, and the intensity of the image are investigated. Particular emphasis is placed on the material parameters necessary for color phase conjugation. The potential of crystals in the tungsten bronze family for providing the needed values for these parameters is explored.

PHOTOREFRACTIVE GRATINGS

One of the more amazing properties of the tungsten bronze family of crystals is their photorefractive behavior. Allow a laser beam with the image of a face imprinted on it to propagate through a crystal (Fig. 1) and the information or details of the image will be stored in the crystal. In order to understand this behavior of photorefractive crystals we begin by first examining a simpler situation shown in Fig. 2. Two monochromatic and identical laser beams interfere at the site of a crystal and produce a spatially modulated intensity pattern. The intensity pattern causes trapped electrons to diffuse to locations which are relatively dark. The diffusion is eventually limited by the restoring force due to the electric field caused by the developing charge separation. At equilibrium diffusion is balanced by this restoring electric field. This electric field can be on the order of thousands of volts per centimeter and consequently, the crystal can become locally distorted via the electro-optic effect. One indication of this distortion is a change in index of refraction. It is this index change which stores the image of the intensity interference pattern in the crystal. There is one difference, however. Since the space charge electric field is 90° phase shifted from the intensity spatial pattern, the resulting index pattern or index grating stored in the crystal is also 90° shifted. While this phase difference may seem to be a minor point, we will soon see that this unique property of photorefractive materials plays an essential role in most photorefractive applications and in the color imaging discussed in this paper.

PHASE CONJUGATION

Having shown that two laser beams can write an index grating in a photorefractive crystal, we may well wonder what evidence there is that the information has indeed been stored. The evidence is shown in Fig. 3. By sending a third laser beam into the crystal, the presence of the index grating can be detected by observing the Bragg deflection of this beam. In this way the third beam is used to read the grating written by the two writing beams. By varying the time delay between the writing of the grating and the reading of the grating, the storage time can be determined. For most photorefractive materials this time is at least on the order of days.

While the writing, storage, and reading are in themselves exciting properties of photorefractive crystals, what is perhaps even more exciting is that the image obtained by reading out the stored information is not only a clear image of the original object, but a phase conjugate image to boot. To see what is meant by the phase conjugate image, we examine Fig. 4. The two writing beams enter from the left side of the crystal and write index gratings. The reading beam enters from the right side and produces the Bragg scattered light, which propagates back to the point A. The key words are "propagates back." This would be the case even if a wavefront distortion, such as that shown in Fig. 5, were present. Put another way, the light originating from point A could be thought of as striking a mirror, as opposed to the crystal, and the Bragg scattered light viewed as light reflected from that mirror. While a regular mirror would be unable to return the incident light back in a counter-propagating direction, the phase conjugate mirror does precisely and uniquely that. Therefore, if in attempting to store an image in a photorefractive crystal the writing beam passes through a wavefront distortion, upon read-out the phase conjugate image undistorts as it passes back through the distorter. A most remarkable property.

BEAM FANNING

As we have discussed, one way to observe the storage of information in a photorefractive crystal is to use a reading beam in order to read out the information. A second indication of information stored in the crystal is to examine the intensity of the two writing beams after they propagate through the crystal. As shown in Fig. 6, the observation is that energy is exchanged from one writing beam to the other resulting in gain for one beam and loss for the other. In order to understand this behavior we must point out two things. First, light scattered from either of the incident beams by the index grating is 90° phase shifted from the incident beams. This shift is due to the fact that when we add up the field at a point P, shown in Fig. 7, due to a plane of oscillating dipoles, the total field at P is 90° lagging the incident or dipole driving field.[1] As a result, light scattered from beam 1 into beam 2 by the stored grating is 90° shifted from beam 2 and light scattered from beam 2 into beam 1 is, likewise, 90° shifted. While the scattered light is indeed shifted in phase, the shift is the same into both beams 1 and 2 and no difference in transmitted intensity is expected. Here, however, a second important point comes into play. Due to the shift between the intensity pattern and the index pattern discussed earlier, a break in this symmetry develops. In Fig. 8 the intensity and index distribution in the crystal are depicted as being in phase. In this case the scattered light from A lags the incident light by 90° for both beams, as discussed. However, in Fig. 9 the intensity and index distribution in the crystal is shown to be 90° shifted. Here the two incident beams arriving at A are not in phase with each other. In fact, they are exactly 90° shifted from one another. Therefore, light scattered from A by beam 1 into beam 2 is shifted forward an additional +90°, while light scattered at A by beam 2 into beam 1 is shifted back, or an additional -90°. When these additional shifts of ±90° are added to the shift due to the plane of dipoles of ±90°, in one case we get a total of 0° degree shift, while in the other case we get a total of 180° shift. Consequently, we have constructive interference in one beam or gain and destructive interference in the other beam or loss, i.e., energy is exchanged from beam 2 to beam 1. As a result, by observing the energy exchange we verify the storage of the index grating.

Actually, this same effect of energy exchange shows up using only one beam incident on a photorefractive crystal. As shown in Fig. 10, light incident on a photorefractive material produces an intense fan of scattered light.[2] The fan is caused by light scattered from crystal defects which cross with the incident beam to produce a weak interference pattern and write a grating. This grating, in turn, causes energy to be exchanged from the strong incident beam to the weak scattered light. The gain experienced by the scattered light then results in the intense fan observed.

PHOTOREFRACTIVE PHASE CONJUGATION

There are several ways in which an image can be stored in a photorefractive crystal and be read out as a phase conjugate image. We will discuss two of them in this paper which will have application for color imaging. These two are: self-pumped phase conjugation[3] and double phase conjugation.[4]

In self-pumped phase conjugation one laser beam enters the crystal causing a stimulated fan pattern to form. As shown in Fig. 11, some of the fanned light hits a corner of the crystal. Light scattered from point A reflects via the crystal corner to point B where it fans and enhances the grating at B. Meanwhile, light scattered from point B reflects via the crystal corner and fans at point A, likewise enhancing the grating at A. The enhanced grating at A then more effectively scatters light at A and writes an even stronger grating at B, and so forth. The process continues as feedback between point A and B causes the intensity between A and B to build up and form an intense loop in the crystal. Meanwhile, when light from A arrives at B, it also reads the grating at B and produces a conjugate image. In the same way light arriving at A from B also produces the conjugate signal.

A second form of self-pumped phase conjugation is shown in Fig. 12. In this case the crystal corner is replaced by a mirror and point A and point B of Fig. 11 become superimposed. As light builds up between point A and the image of A through the mirror or point B, the phase conjugate signal again appears.

A second method for producing a phase conjugate image is called double phase conjugation and is shown in Fig. 13. In this case two separate lasers are used. Light from one laser produces the fan at point A, while light from the second laser produces the fan at point B. Aside from this difference, the process is then the same as that described in Figs. 11 and 12. That is, light fanned from A arrives at B and enhances the fan grating at B, while light from B enhances the grating at A. Meanwhile, light from A also reads the grating at B producing the conjugate image along beam 2 and light from B reads the grating at A producing the conjugate image along beam 1. In this way two separate conjugate images are produced without crosstalk between them.

MULTI-COLORED FAN

Our discussion to this point has been to prepare us to understand color imaging and color phase conjugation in photorefractive crystals. Suppose we examine the photorefractive effects just discussed using a multi-colored argon-ion laser. That is, a laser oscillating at several discrete lines simultaneously. For example, would we observe the intense fan pattern shown in Fig. 14? The answer is yes we do, but the intensity is orders of magnitude weaker. Apparently, light at one wavelength tends to erase the gratings written at a second wavelength and vice versa. As a result, the gratings are much less pronounced and the stimulated fan pattern is much weaker than would be expected based on the scattering which occurs for the case of a single incident laser wavelength.

As a result of the considerably weaker fan pattern and competition between different colored loop or bridging patterns, it is difficult to observe self-pumping using the crystal corner reflection, as well as double phase conjugation using two separate multi-colored laser beams.

RAINBOW EFFECT

Recently, we have observed an effect which overcomes the difficulty of erasure between colors when a multi-colored laser source is used to write gratings. In particular, for the arrangement of Fig. 15, when a multi-colored laser beam enters the photorefractive crystal, we have observed that each color enhances one another and a multi-colored intense rainbow pattern is observed. The explanation for this pattern can be seen after examining Fig. 16. Incident light of one wavelength along direction k^i produces scattered light which writes a grating of wave vector k^g. This results in a fan of scattered light with different values of grating wavelength or periodicity indicated by different values of k^g. When more than one color is present in the incident beam, both produce their own fan patterns which may interfere with one another. At one special angle, however, shown in Fig. 17, the grating wavelength for incident blue light, for example, will be the same as for the slightly dispersed incident green light. That is, at this special angle the incident blue light is Bragg matched to the grating written by the incident green light and the incident green light is Bragg matched to the grating written by the incident blue light, i.e.,

$$k_B^g = k_G^g$$

As a result of the Bragg matching between colors, the grating written by blue light scatters green light, which therefore writes a stronger grating with the incident green light. The strong green grating then scatters more blue light, which likewise writes a stronger grating with incident blue light. In this manner feedback is provided between the blue and green fanned grating and an intense ring pattern forms. The pattern is in the form of a ring due to the fact that this argument can be made for the incident beam and any fanned beam lying in a cone, as shown in Fig. 18. As indicated in Fig. 19, this phenomenon occurs in the tungsten bronze crystal strontium barium niobate only for negative incident angles. The reason for this behavior is shown in Fig. 20. For the negative angle in Fig. 20a we can find a direction in the crystal where the grating written by blue light is identical in magnitude and direction to that written by green light. However, for the positive angle in Fig. 20b there is no angle for which the fanned grating written for blue light is identical to the fanned grating written by green light. As a result, for this orientation each color tends to erase the grating written by other colors as opposed to reinforcing them.

COLOR IMAGING AND PHASE CONJUGATION

Color imaging was first reported in reference 6. In this case an external mirror was used, as in Fig. 12, to provide the feedback needed for self-pumped phase conjugation to occur. While these results were impressive, they did point out the existence of competition and crosstalk between colors. By using the geometry of Fig. 21 and back reflecting the rainbow pattern into the crystal, color conjugation is obtained with high resolution and without competition and crosstalk between colors. In this case no conjugate is obtained with only one color present. Rather, multi-colors are <u>required</u> in order to observe a conjugation signal, or store and read an image of any type. As seen in the conjugate color image in Fig. 22, the separation between the smallest two horizontal lines is about 1/30 of a millimeter corresponding to a resolution of better than 30 lines per millimeter. Moreover, the conjugate was produced in four colors which corresponded to the four most intense lines of the argon-ion laser. The three most intense color images were examined using interference filters to select out only one laser wavelength. These images are shown in Fig. 23. The input object is shown in Fig. 24 for comparison.

A second method for producing color phase conjugation is shown in fig. 25. In this case double phase conjugation was made to occur in color, giving results similar to that observed in Figs. 22 and 23. However, in this case the conjugate signal appears along both input beams.

The response time for the formation of the color image for both methods of color conjugation is shown in Fig. 26. Here we see that the response time for the formation of the multi-colored rainbow pattern is the same as the single colored beam fanning time and that the time for image formation is only several times longer. In addition the behavior of the conjugate reflectivity as a function of incident angle is given in Fig. 27. As can be seen reflectivities of greater than 10% were easy to achieve.

For completeness we point out a third method for producing color imaging or phase conjugation. The technique is shown in Fig. 28. In this approach we have split the incident beam spatially into separate colors. As a result, each beam propagates into the crystal in a different spatial plane, making it possible for self-pumped phase conjugation, using only the crystal corner, to be observed. That is, each color can form a loop pattern which does not interfere with the loop formation of another color due to the spatial plane separation. While this technique leads to clear color images, similar to those shown in Figs. 22 and 23, its response time was considerably slower.

CONCLUSIONS

We have demonstrated several techniques for producing color imaging in tungsten bronze crystals. Our observations were carried out in SBN:60,

SBN:75, BSKNN II and BSKNN III using various dopant materials. Color images of a centimeter in size were stored in the crystal with a resolution exceeding 30 lines per millimeter. The phase conjugate image was produced with a reflectivity exceeding 10% and with a response time only several times longer than the crystal beam fanning response time. It was discovered that in order to produce good color images the most important crystal parameter was its spectral response. Because of the open structure of the tungsten bronze material, it has been possible to engineer crystals with a broad response over the visible spectrum. As a result, the tungsten bronze crystals are particularly well suited to color imaging applications. Future research will include the use of a kryton-ion laser in order to produce a broader spectrum of light and attempts to produce color 3D holograms.

REFERENCES

1. R.P. Feynman, "Diffraction," in The Feynman Lectures on Physics, p. 30-10, Addison-Wesley Publishing Co., Reading, Mass., 1966.

2. J. Feinberg, "Asymmetric Self-defocusing of an Optical Beam from the Photorefractive Effect," Journal of the Optical Society of America, vol. 72, pp. 46-51.

3. J. Feinberg, "Self-pumped, Continuous-wave, Phase Conjugator Using Internal Reflection," Optics Letters, vol. 7, pp. 486-488, 1982.

4. S. Weiss, S. Sternklar, and B. Fischer, "Double Phase-conjugate Mirror: Analysis, Demonstration, and Applications," Optics Letters, vol. 12, p. 114 1987; S. Weiss, S. Sternklar, M. Segev, and B. Fischer, "Mach-Zender Interferometer with Multimode Fibers Using the Double Phase-conjugate Mirror," Applied Optics, vol. 25, p. 4518, 1986; S. Weiss, S. Sternklar, and B. Fischer, "Optical Information Processing with the Double Phase-conjugate Mirror," Optics Engineering, vol. 26, p. 423, 1987; R.W. Eason and A.M.C. Smout, "Bistability and Noncommutative Behavior of Multiple-beam Self-pulsing and Self-Pumping in BaTiO3," Optics Letters, vol. 12, p. 51, 1987; A.M.C. Smout and R.W. Eason, "Analysis of Mutually Incoherent beam Coupling in BaTiO3," Optics Letters, vol. 12, p. 498, 1987; M.D Ewbank, "Mechanism for Photorefractive Phase Conjugation Using Incoherent Beams," Optics Letters, vol. 13, p. 47, 1988; "Incoherent Beams Sharing Photorefractive Holograms," in Digest of Topical Meeting on Photorefractive Materials, Effects, and Devices (Optical Society of America, Washington, D.C., 1987), p. 179; E.J. Sharp, W.W. Clark, III, M.J. Miller, G.L. Wood, B.D. Monson, G.J. Salamo, and R.R. Neurgaonkar, "Double Phase Conjugation in Tungsten Bronze Crystals," to be published in Applied Optics.

5. G.J. Salamo, M.J. Miller, W.W. Clark, III, G.L. Wood, E.J. Sharp, and

R.R. Neurgaonkar, "Photorefractive Rainbows," Applied Optics, vol. 27, p. 4356, 1988.

6. M. Cronin-Golomb, S. Kwong, and A. Yariv, "Multi-color Passive (Self-pumped) Phase Conjugation," Applied Physics Letters, vol. 44, p. 727, 1984.

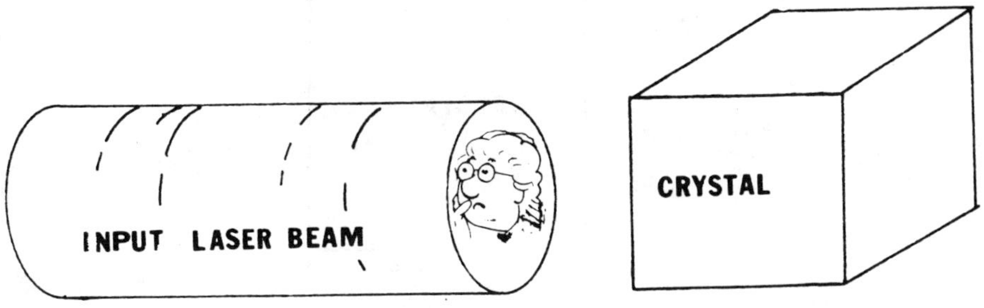

Fig.1. IMAGE STORAGE

FIG. 1 Incident laser beam with scene imprinted on it is directed into the photorefractive crystal. The scene is then stored in the crystal bulk.

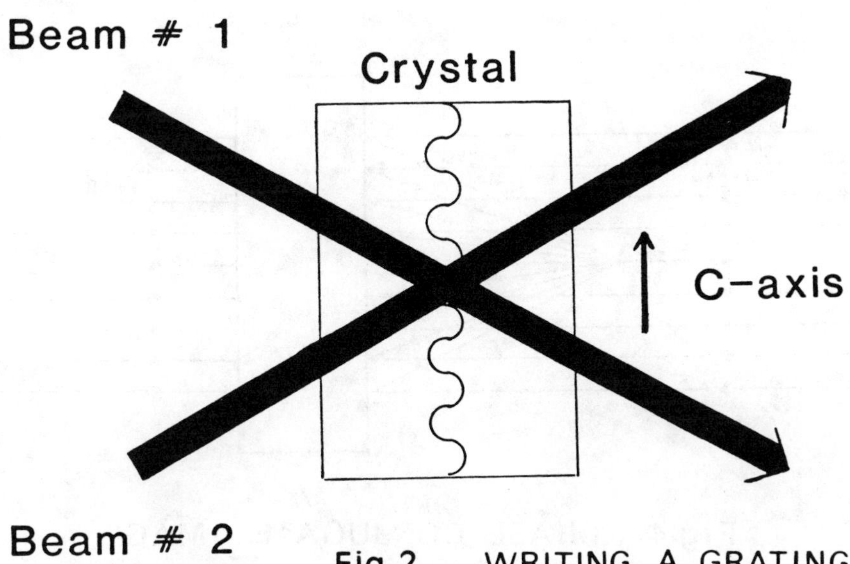

Fig.2. WRITING A GRATING

FIG. 2 Two laser beams intersect inside the photorefractive crystal and produce a modulated intensity distribution. The intensity pattern then results in a corresponding index pattern.

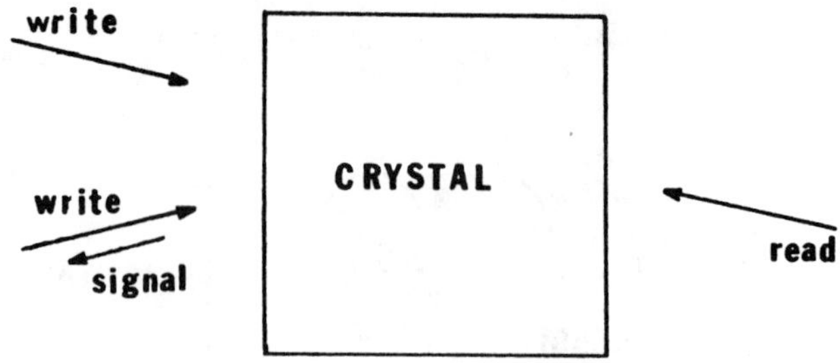

Fig.3. READING A GRATING

FIG. 3 Two laser beams are used to write an index grating inside the photorefractive crystal, while a third laser beam called the reading beam is used to detect the index grating. The signal beam is produced by Bragg diffraction of the reading beam off of the index grating.

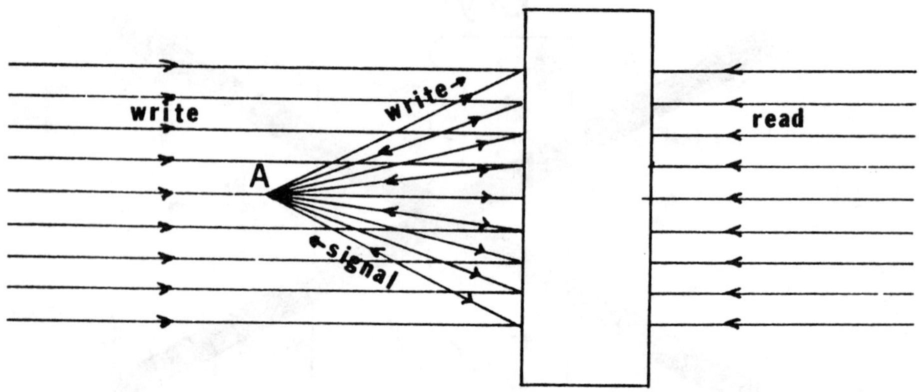

Fig.4. PHASE CONJUGATE IMAGE

FIG. 4 Two incident beams from the left meet and write gratings in the crystal, while the reading beam, incident from the right, produces the signal which reconstructs the point source image.

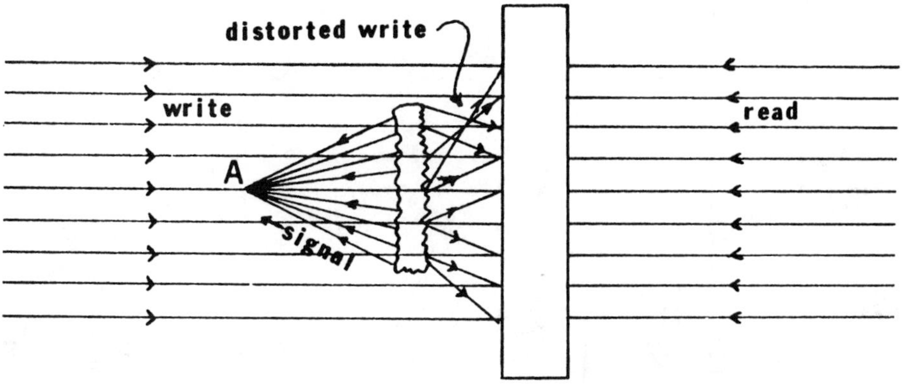

Fig.5. PHASE CONJUGATE CORRECTED IMAGE

FIG. 5 Two incident beams from the left enter the crystal and write gratings even though one of the beams passes through an optical distortion. The reading beam enters from the right and generates the phase conjugate signal which reconstructs the original point object after it passes through the optical distorter.

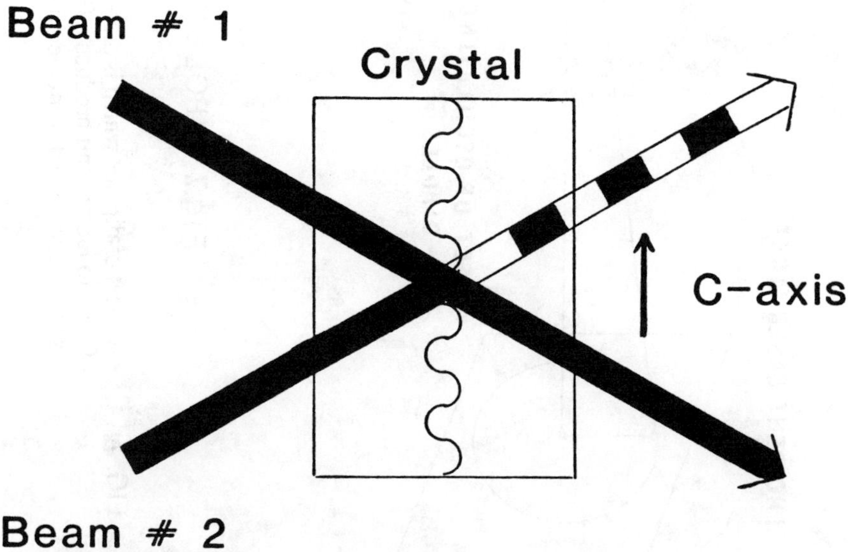

Fig.6. TWO BEAM COUPLING

FIG. 6 Two beams enter a photorefractive crystal and write a grating. Once the grating is written, energy is exchanged from beam #2 into beam #1.

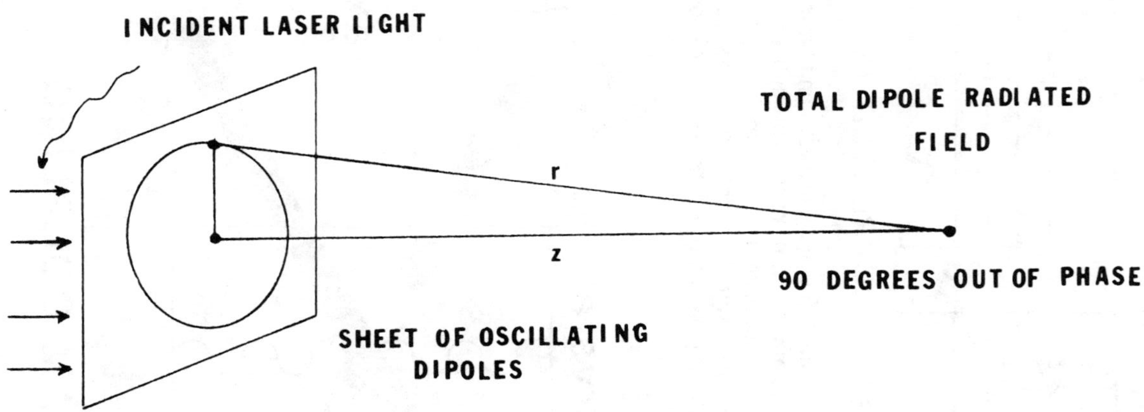

Fig.7. DIPOLE FIELD

FIG. 7 A sheet of dipoles which is driven by incident laser light is found to radiate and produce a field along the Z direction which is 90° phase shifted from the driving field.

BEAM FANNING CONCEPT

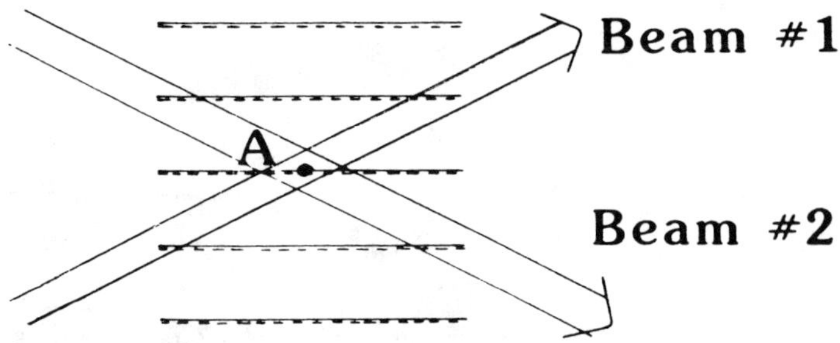

Intensity Peaks Overlap Index Peaks

FIG. 8 Two incident beams write a grating inside the crystal. Light scattered at A from beam #1 into beam #2 is 90° shifted from #2, while light scattered from beam #2 into #1 is also 90° shifted from #1.

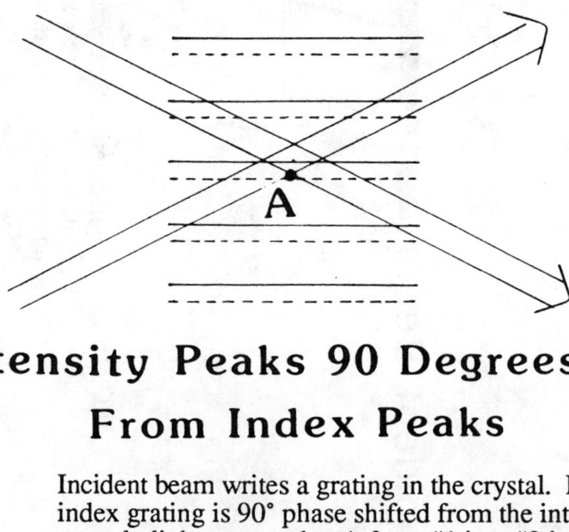

Intensity Peaks 90 Degrees Shifted From Index Peaks

FIG. 9 Incident beam writes a grating in the crystal. In this case the index grating is 90° phase shifted from the intensity pattern. As a result, light scattered at A from #1 into #2 is 90° leading the phase of #2, while light scattered from #2 into #1 is 90° lagging the phase of #1. When this additional shift is added to the 90° shift due to the sheet of dipole radiation, the result is that light scattered into #1 is 180° lagging the field of #1, while light scattered into #2 is in phase with #2.

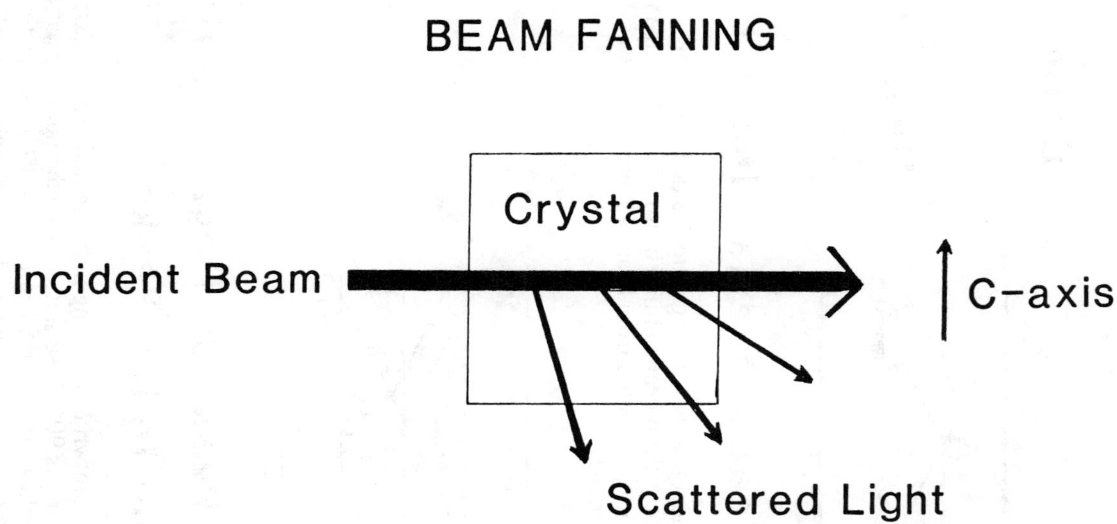

FIG. 10 Light entering the crystal is scattered off of crystal imperfections. The scattered light overlaps the incident beam and gratings are written and energy is transferred from the incident beam into the scattered light, causing an intense fan pattern to form.

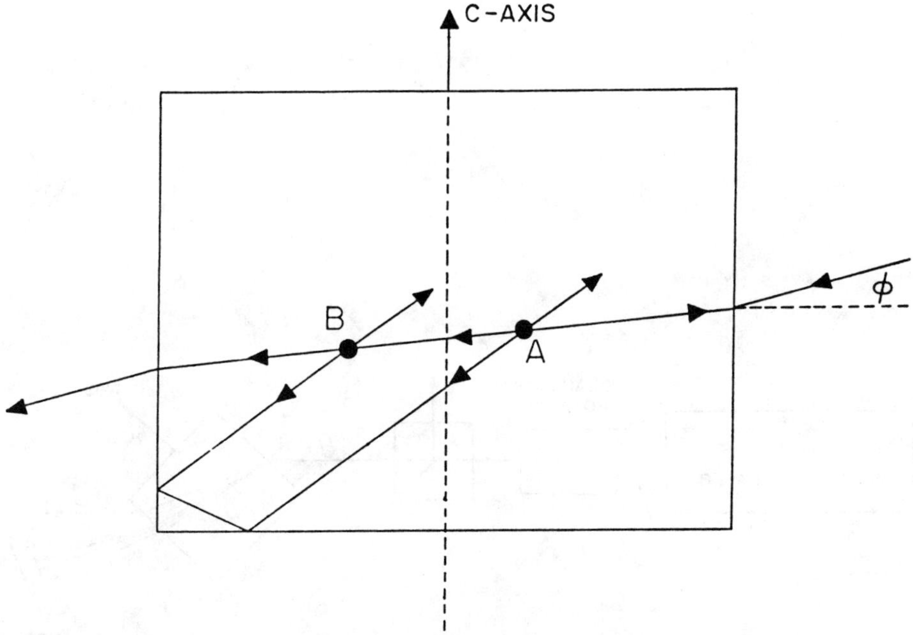

Fig.11. SELF-PUMPING

FIG. 11 Light entering the crystal undergoes fanning. Some of the fan light from A reflects around the crystal corner to B where its fan gratings reinforce the gratings at A. Likewise, light fanned at B reflects via the crystal corner to A where it fans and forms gratings which reinforce the gratings at A. This process continues as an intense loop is formed inside the crystal. Meanwhile, light from A reads the grating at B and produces a conjugate wave, while light from B reads the grating at A and also produces a phase conjugate wave.

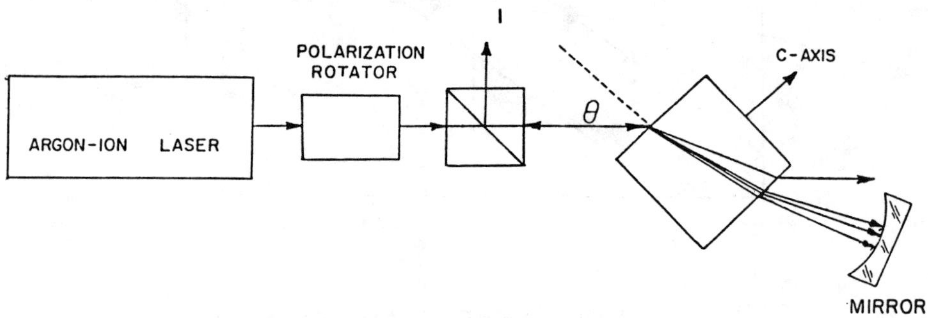

Fig. 12. SELF-PUMPING

FIG. 12 Laser light enters the photorefractive crystal and fans. The fan light is reflected back into the crystal and its fan gratings reinforce the grating already in the crystal. This process continues until a strong beam is observed between the crystal and mirror. The light reflected into the crystal also reads the crystal gratings and produces a conjugate wave.

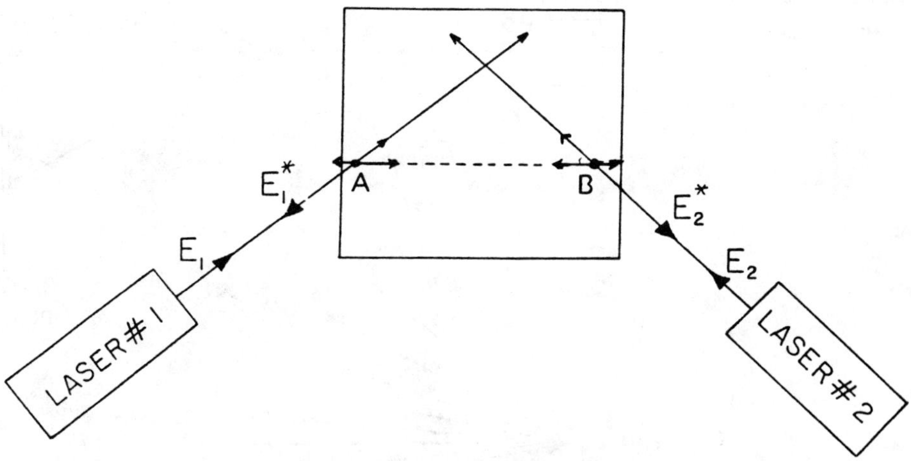

Fig.13. DOUBLE PHASE CONJUGATION

FIG. 13 Two beams from different laser sources enter the crystal. At point A fanning occurs and light is fanned to point B where it writes gratings which reinforce the gratings at B. Likewise, light which fans at B arrives at A where it fans and reinforces the gratings at A. This process continues until a strong beam appears inside the crystal between A and B. Meanwhile, light from A reads the gratings at B, while light from B reads the gratings at A. In this way the phase conjugate of both beams #1 and #2 are produced.

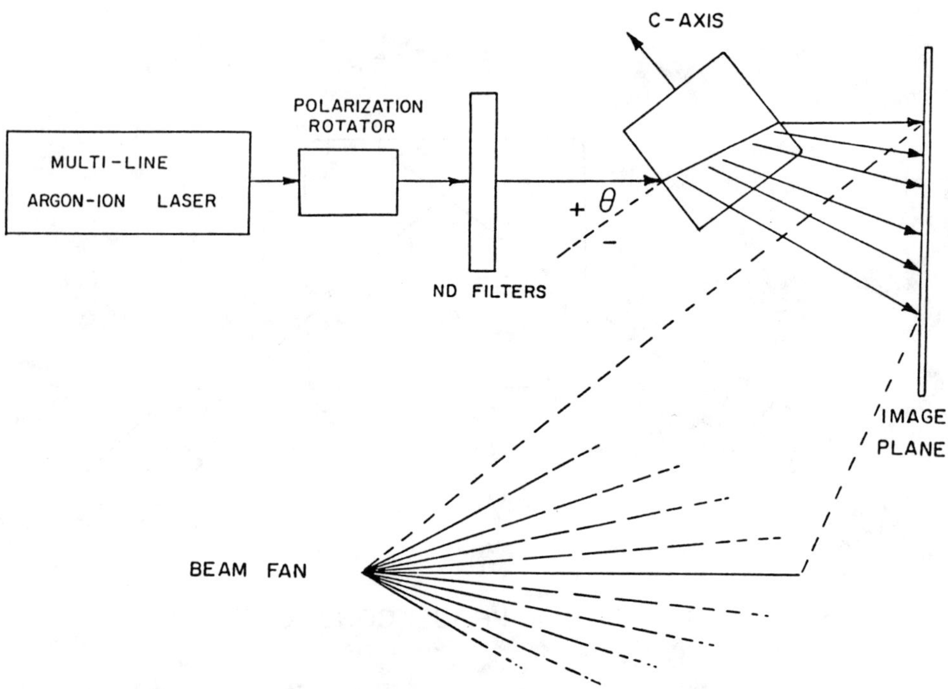

Fig.14. COLOR FANNING

FIG. 14 Multi-colored laser light enters a photorefractive crystal. While each color fans, it is also found that one color tends to erase the other. As a result, the total fan intensity is greatly reduced over what it would be if it was a single color that was incident on the crystal. About 99% of the incident light is fanned using a single laser line, while only 20% is fanned using multi-lines.

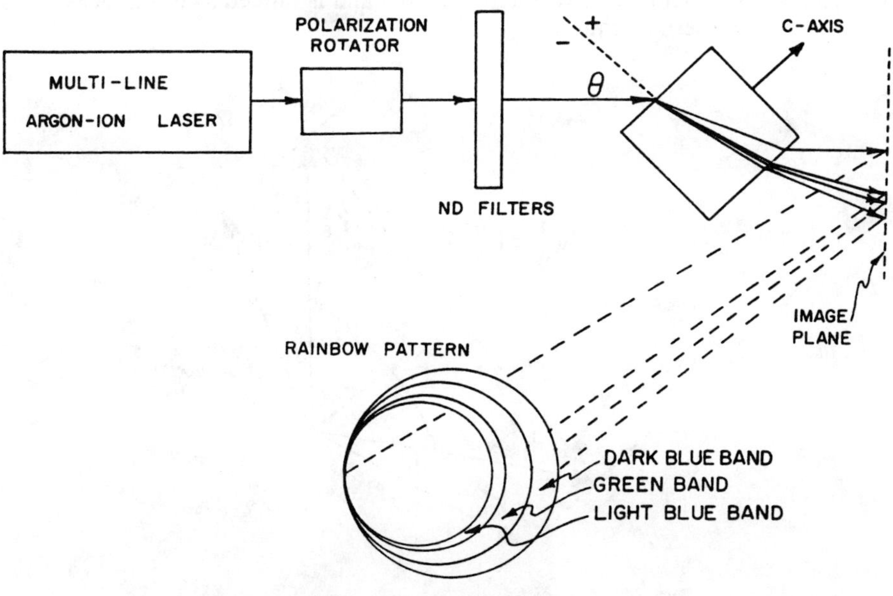

Fig.15. COLOR RAINBOW

FIG. 15 A multi-colored laser beam enters a crystal. An intense multi-colored ring pattern forms in a time corresponding to the photorefractive response time.

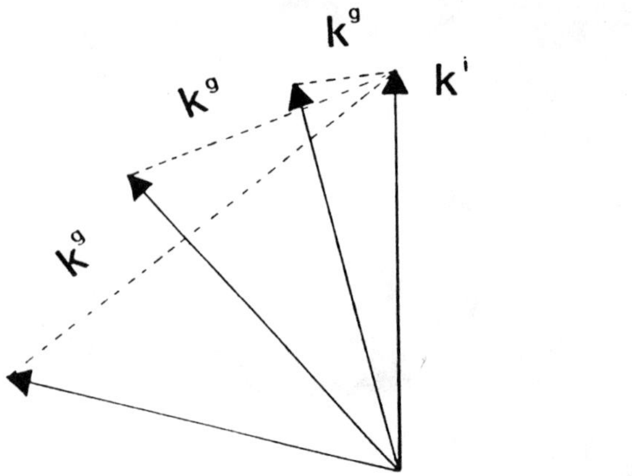

Fig. 16. BEAM FANNING

FIG. 16 An incident wave enters a crystal and is fanned as it produces numerous gratings.

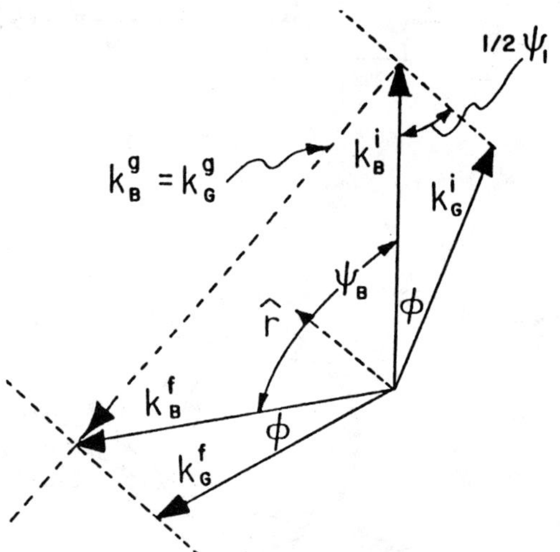

Fig.17. COLOR BRAGG DIFFRACTION

FIG. 17 When two colors enter the crystal dispersion separates the waves inside the crystal. Each color fans independently of each other. However, at one particular angle the grating written by blue light is reinforced by the grating written by green light and the scattering process becomes stimulated.

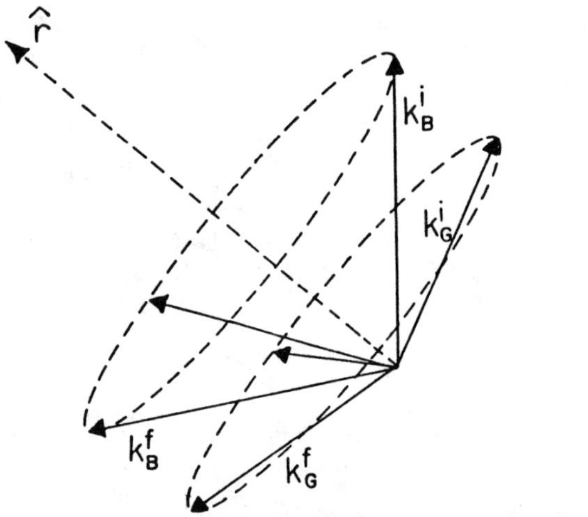

Fig.18. COLOR BRAGG DIFFRACTION AROUND THE CONE

FIG. 18 The argument made in the previous figure can be extended to a set of fanned wavevectors which lie on a cone around the vector r.

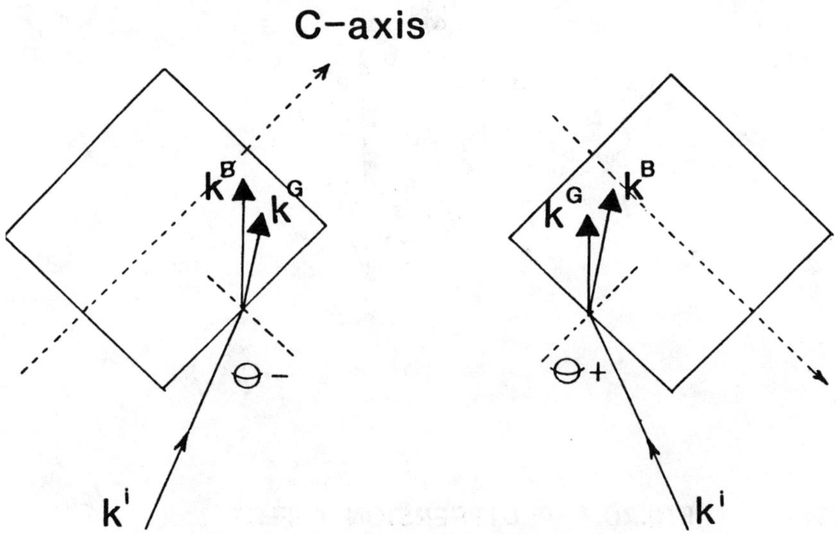

Fig.19. DISPERSION EFFECT

FIG. 19 Light is incident on the crystal with either a positive or negative angle. As shown, dispersion is different in each case.

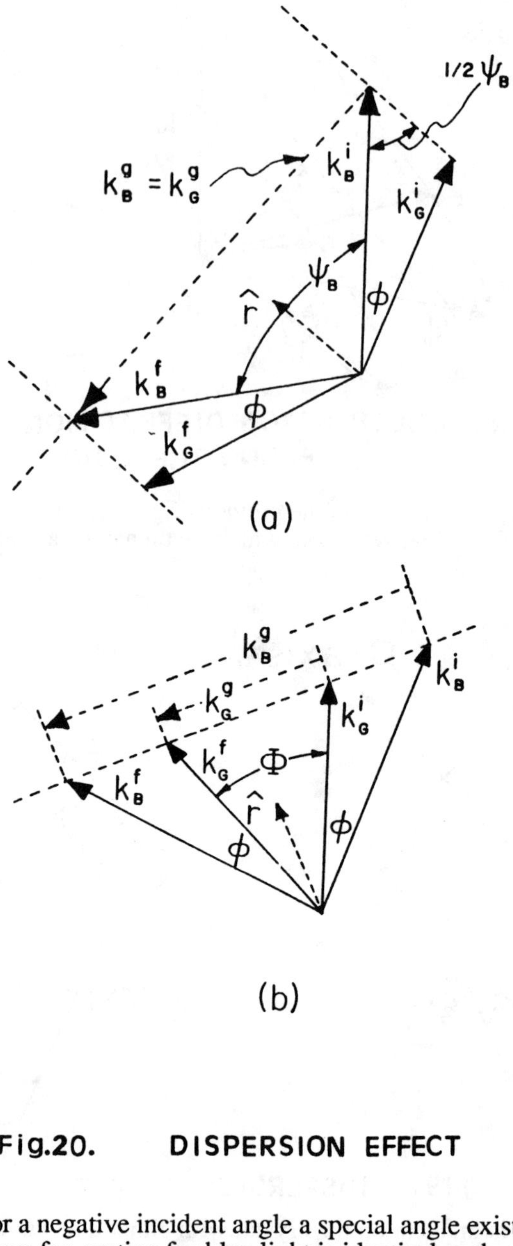

Fig.20. DISPERSION EFFECT

FIG. 20 For a negative incident angle a special angle exists where the beam fan grating for blue light is identical as that for green light.

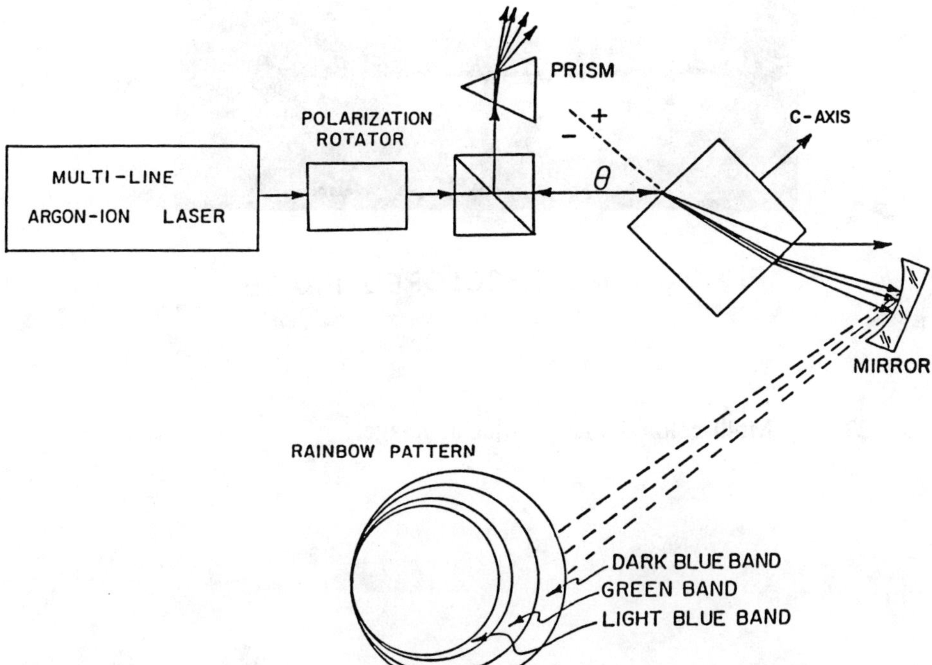

Fig. 21. RAINBOW CONJUGATION

FIG. 21 Using a mirror the rainbow pattern is back reflected into the crystal and a multi-colored conjugate appears.

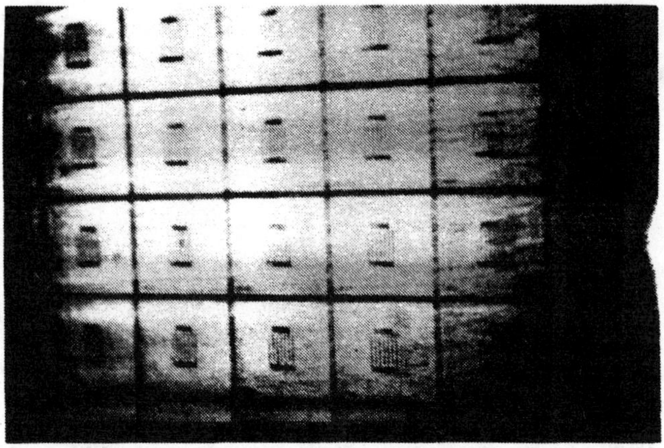

Fig.22. MULTI-COLORED IMAGE

FIG. 22 Multi-colored phase conjugate image.

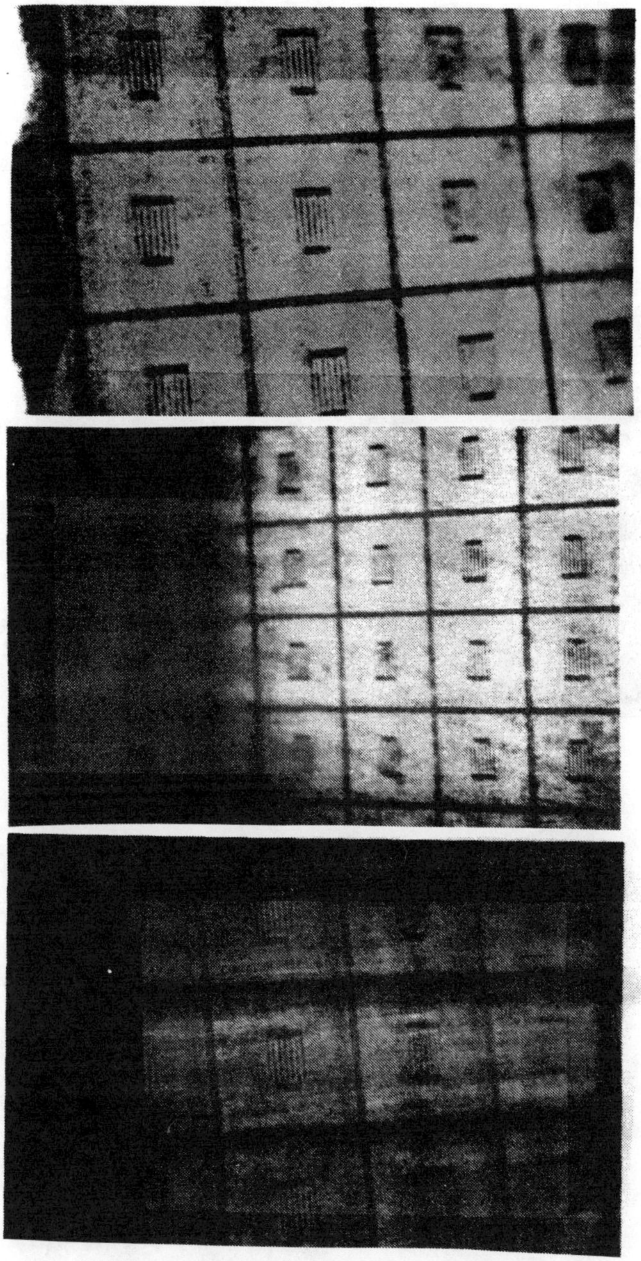

Fig. 23 Filtered Images

FIG. 23 Multi-colored image after dispersion.

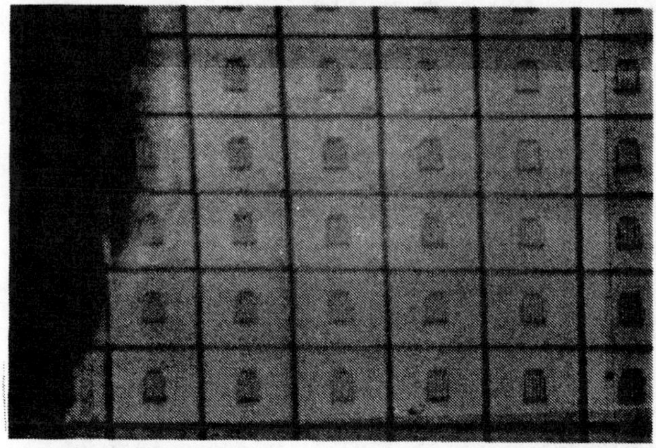

Fig.24. INPUT IMAGE

FIG. 24 Input scene.

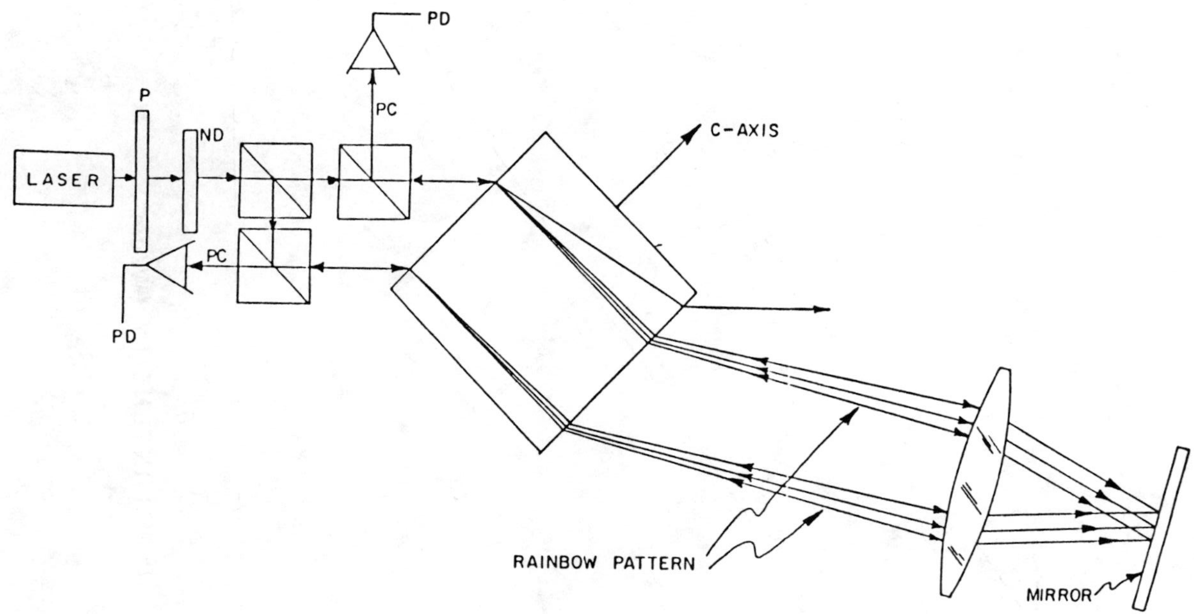

Fig 25 RAINBOW DOUBLE PHASE CONJUGATION

FIG. 25 Using two multicolored beams two rainbow patterns are produced and back reflected one into the other as a color double conjugate is formed.

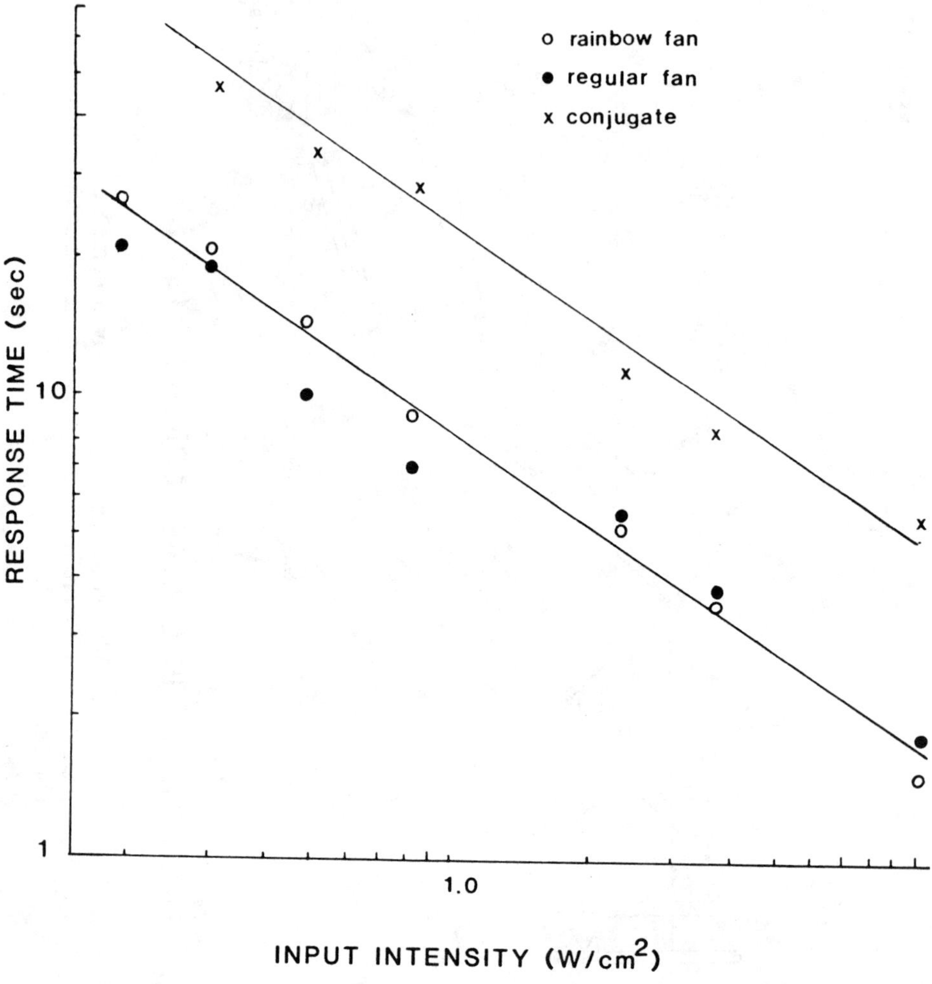

FIG. 26 Time response for the multicolored beam fan and phase conjugate compared to the normal single color fan.

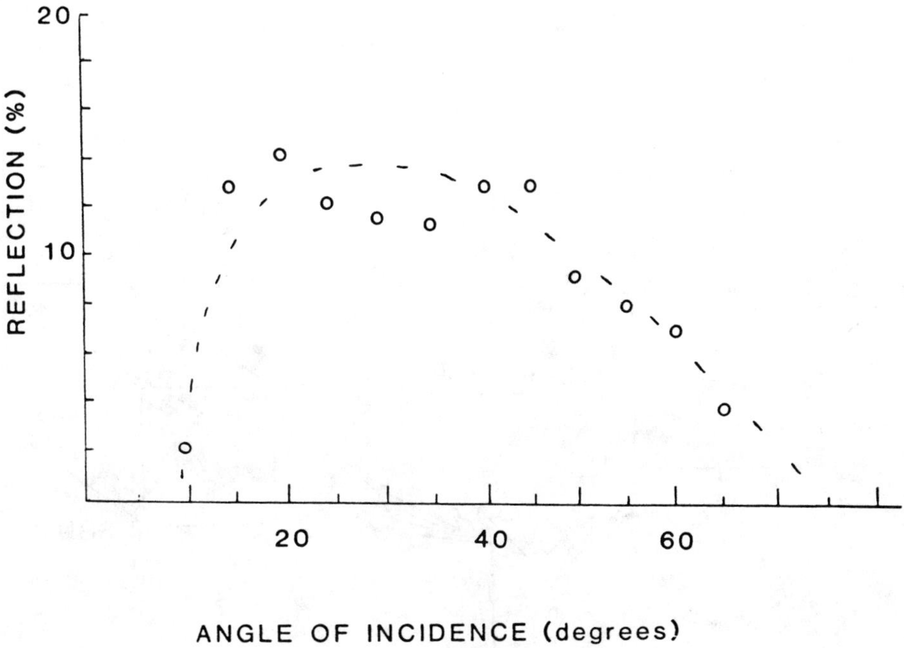

FIG. 27 Angle behavior of the color conjugate. Data shows that the conjugate reflectivity for the configuration of fig. 21 is 10% for a broad range of incident angles.

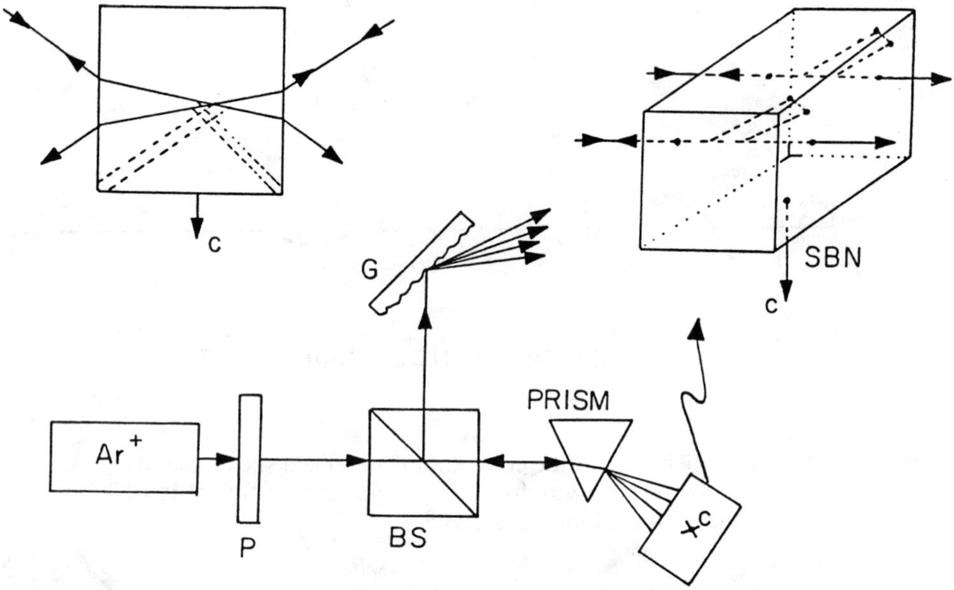

Fig. 28. COLOR SELF–PUMPING

FIG. 28 Incident multi-colored light is dispersed before entering the crystal in order to cause a self-pumped loop to form for each color independently.

TWO VIDEO DISPLAYS USING PLZT LIGHT SHUTTERS

Kanji Murano
Sony Corporation
Corporate Research Laboratories
6-7-35 Kitashinagawa, Shinagawa-ku
Tokyo, 141 Japan

ABSTRACT

This paper reports two PLZT displays:
(1) A PLZT color viewfinder in which tri-color separation PLZT light shutters are used to convert the B/W image of a CRT to a color image.
(2) A PLZT video projector in which a beam from a Xenon light source passes through the PLZT linear array and the resulting horizontally modulated beam is scanned vertically by a movable mirror and projected on the screen.

INTRODUCTION

A PLZT light shutter has the advantages of a higher speed response than a liquid crystal light shutter and a lower operation voltage compared to that of the other light shutters utilizing electro-optic single crystals. PLZT light shutters have already been used as flash blindness prevention shutters, viewers in a stereoscopic television system and optical printer shutters.

PLZT light shutters are used in different ways depending on the video display system. The horizontal lines of a display are scanned sequentially to display a picture. TV system have different scanning rate and number of horizontal lines. The US and Japan use the NTSC system in which the horizontal scanning time is 63.5 μs including 10μs retrace time or blanking time, and a picture consists of 525 horizontal lines. To eliminate annoying flicker, the interlaced horizontal scanning pattern which divides the horizontal lines into two groups called fields. In the first field half of the 525 lines are scanned on the screen in 16.7 ms including 1 ms vertical retrace time, and in the second field the remaining lines which are between the first field lines are scanned.

A stereoscopic television system uses this two field system. A stereo picture is produced by a two-field image, one field is from the left camera and the other from the right camera. With a PLZT light shutter goggle the eye sees the left field image and the right eye the right field image. In this case the PLZT light shutter must be switched in the vertical blanking time of 1 ms.

We have reported two video displays using PLZT light shutters. One is a battery-operated color viewfinder for a video camera using a tri-color separation PLZT light shutter and a black/white (B/W) cathode ray tube (CRT), the other is a light valve video projector using a PLZT light shutter array. The PLZT viewfinder utilizes line sequential scanning system and converts the B/W CRT image to a color image using the PLZT light shutters. The PLZT projector uses the principle of the horizontal scanning system and a projected line corresponds to a light modulated PLZT shutter array.

This paper reports our developmental work and discusses the problems of application PLZT light shutters to video displays.

A PLZT COLOR VIEWFINDER [1, 2]

At present an electronic viewfinder for a portable color video camera shows a black and white image because a color CRT of less than 1.5 inches has a complex structure which is difficult to make. However, the need for an electronic viewfinder showing a color TV image is obvious. Using PLZT light shutters and a 1.5-inch B/W CRT, a simple line-sequential color reproducing apparatus was developed, which was found to give an excellent color TV image. The system configuration of the electronic color viewfinder is shown in Fig.1. The system configuration of the light shutter is shown in Fig.2. A timing diagram of this system is shown in Fig.3.

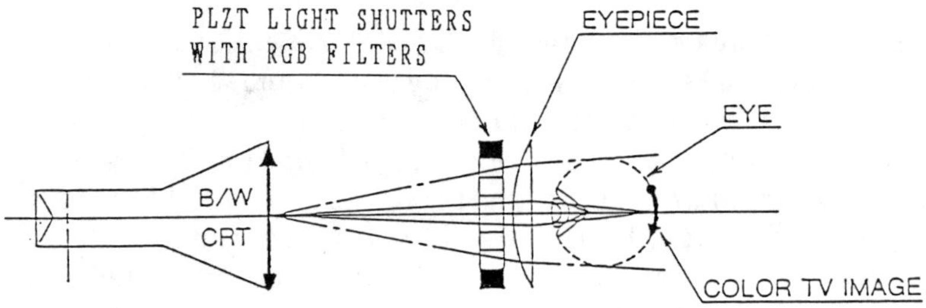

Fig. 1. System Configuration.

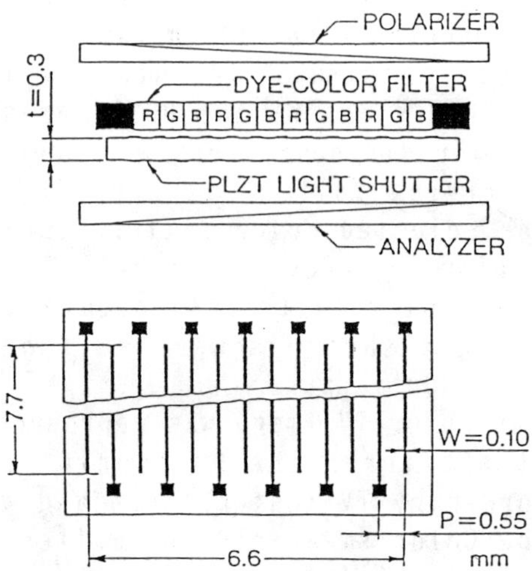

Fig. 2. Configuration of Light Shutters.

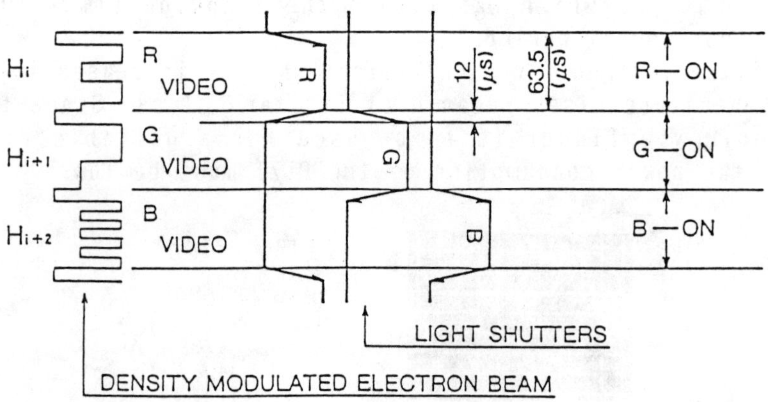

Fig. 3. Timing Diagram of the System.

The electron beam of the CRT is density-modulated by the color video signals. The light shutter opposite a Red color filter is open when a Red video signal is called for. In order to reproduce a line-sequential color TV image, the response time of PLZT light shutters has to be less than the blanking time (10 μs) during a horizontal period.

The light path from a 1.5 inch CRT must be switched and the RGB components from a 1.5 inch CRT must be selected to reproduce the color TV image. PLZT light shutters were used to switch the light path. Dye-color filters were used to select the RGB components. A shutter array with 12 stripes was prepared. The array consisted of interdigital electrodes glazed with Ag on both sides of the PLZT, as shown in Fig. 2. Four sets of RGB color stripe filters were put on the shutter array. When the driving voltage is supplied to electrodes which opposite the selected color filter, the light shutters are open and the color components of the 1.5-inch CRT are then selected by the dye-color filter. To obtain a fine color TV image, the PLZT light shutters were switched at 5.25 KHz square waves of a 1/3 duty ratio.

A high resolution color TV image was reproduced by the high-speed PLZT light shutters. The horizontal resolution was measured by the ordinary TV test pattern and was found to be 250 TV lines/inch. Color smear did not matter as long as the eyepiece was placed properly against the eye. Figure 4 shows the response of the PLZT light shutters. The response time was less than 10 μs, which was within the blanking time (10μs) during a horizontal period.

AC driving voltage for a PLZT light shutter increases as the PLZT temperature rises because of its tan δ loss. Since this electronic viewfinder is to be used for a portable video camera, the power consumption of the PLZT must be low.

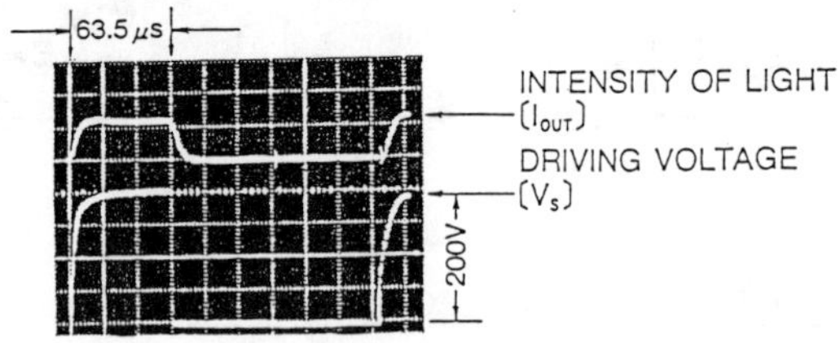

Fig. 4. Response of PLZT Light Shutters.

A PLZT with low power consumption must have a low dielectric constant ε_s, low tanδ, and a large transverse quadratic electro-optic coefficient R at the operating temperature. 9/65/35 PLZT is well known as an useful shutter composition at temperatures higher than -20℃, however it is known that 8/65/35 PLZT can be used as low power shutter composition at temperatures higher than 20 ℃. In order to investigate PLZT power requirements at temperatures higher than 24℃, we prepared x/65/35 compositions and measured the temperature dependence of power consumption. The halfwave voltage or driving voltage of the PLZT light shutter was measured under 5.25 KHz AC applied voltage at various temperatures using the experimental setup shown in Fig.5. Halfwave voltage was determined by the first transmittance peak voltage.

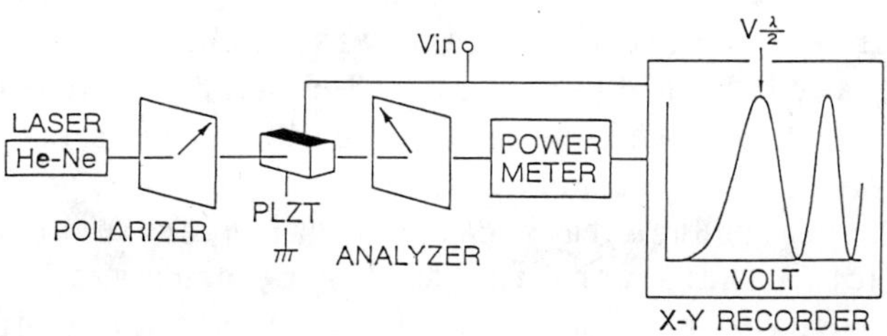

Fig. 5. Experimental Setup for Halfwave Voltage Measurements.

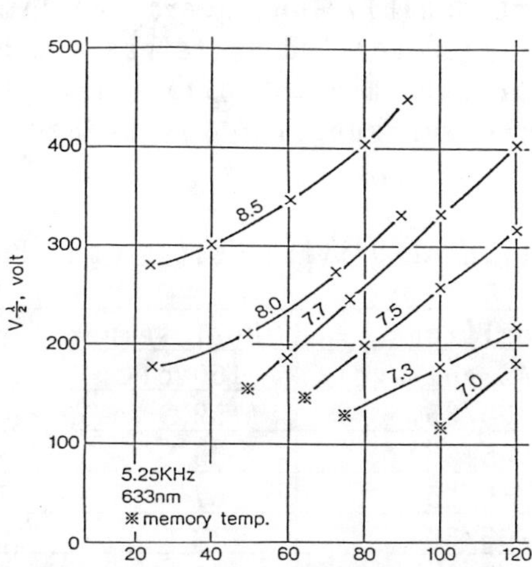

Fig. 6. Halfwave Voltage on the Light Shutter as a function of Temperature for PLZT x/65/35.

Temperature dependence of the AC halfwave voltage of the x/65/35 PLZT light shutter in Fig.2 is given in Fig.6. Table 1 shows the relation between the applied AC voltage and the PLZT temperature for x/65/35 PLZT compositions. If a PLZT shutter using 8.0/65/35 is held at 24 ℃, halfwave voltage is 170 volts(Fig.6). However, if the temperature is not held at 24 ℃, the PLZT temperature increases gradually up to around 70℃ because of tan δ loss. (See Table 1), where halfwave voltage is 260 volts (See Fig.6). It is diffcult to determine halfwave voltage if the PLZT temperature is not kept at constant. The PLZT temperature varies with the environmental temperature, structure of shutter and driving voltage. From our investigations we conclude that a shutter with a 7.3/65/35 composition is best suited to operate at a low driving voltage at room temperature. The operating power required by this shutter at room temperature was 2.5W. For comparison, when we operated a shutter with a 8.5/65/35 composition the operating voltage and power were 440 volts and 10.6W, respectively, at room temperature.

In order to introduce this color viewfinder to actual video camera products, there were the following problems.

① The total power consumption of the portable video camera system was around 8W. Additional 2.5W power for PLZT was too large to use for reasonably long operating time.

② The driving voltage over 100 volts was too high to use in a video camera. To produce a high voltage from a 9-volt battery, a large switching power supplier was needed, which made the camera bigger.

Table 1. The Relation of Driving Voltage and PLZT Temperature.

Compositions	PLZT temperature [℃]	
(Applied AC voltage)	150volt	200volt
8.5/65/35	52℃	67℃
8.0/65/35	65	80
7.7/65/35	67	77
7.5/65/35	77	91
7.3/65/35	86	94
7.0/65/35	98	110

A VIDEO PROJECTOR USING A PLZT LIGHT SHUTTER ARRAY [3)]

There have been many efforts to develop large-size video displays. A CRT may be the best display from the point of view of reliablity, cost, color quality and resolution, but if a display measuring more than 40 inches is required, a video projection system may provide the best answer. A light-valve projector using a Xenon light source and video light valves has high brightness and high resolution. This type of projector is similar to a slide projector, but it uses write and erasable video film in place of a slide.

We have developed a light-valve projector using PLZT light shutters. A linear PLZT light shutter array of 64 light valves is used as a light modulator and a horizontal light scanner. To make a two-dimensional image, a slow-speed movable mirror is used for vertical scanning.

A beam from the Xenon light source passes through the PLZT light shutter array and the resulting modulated beam is scanned vertically by a movable mirror and projected on the screen, as in Fig. 7. The PLZT light shutters in the array are modulated with analog signals sampled from the video signal. The PLZT shutters have a memory effect similar to a capacitor. Each shutter holds its signal for a horizontal scanning period of 63.5 μs, maintaining its light intensity.

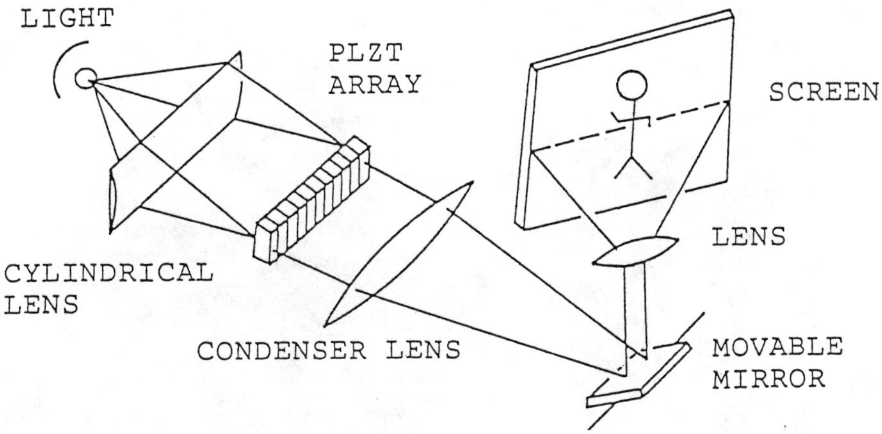

Fig. 7. PLZT Projector System.

The PLZT shutter array, the central device of a PLZT projector, must meet the following requirements.[4, 5]

(1) The shutters should have a compact, high-density arrangement and it should be easy to attach the many leads from each shutter.

(2) Crosstalk among shutters caused by stray capacity should be eliminated.

(3) They should have a low driving voltage and a high contrast ratio.

We have developed a new PLZT light shutter array using parallel-plate electrodes as in a capacitor. (See Fig. 8.) A PLZT light shutter with parallel-electrodes has a low driving voltage and a high contrast ratio because the electric field inside the PLZT is uniform. If many PLZT light shutters are separated from each other and arranged in line, we can get an ideal shutter array free from crosstalk. Problems in the arrangement of the small PLZTs, in making the electrodes and in wiring have made it difficult in the past to produce this type of PLZT array, but we think we have solved these problems.

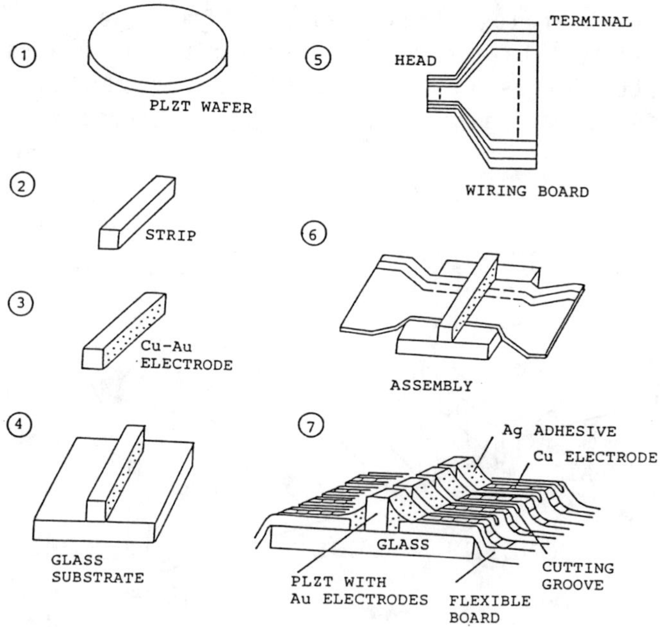

Fig. 8. PLZT Array Construction Sequence.

A projector using the PLZT shutter array shown in Fig. 9 was tested. The width of the beam passing through the PLZT array had to be condensed by 1/3 by the L3 condenser lens because the movable mirror which has a response time of 1 ms, the flyback time of vertical scanning, was too small to adapt the full-size beam. We used only commercially available parts and devices to make the projector.

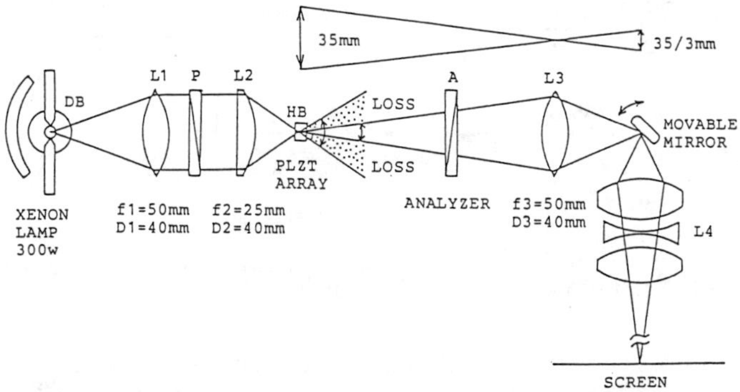

Fig. 9. Configuration of PLZT Projector.

Figure 10 shows the circuit of the video projector. The video signal (1V $_{pp}$) was amplified up to 70V $_{max}$, which was the maximum half-wave voltage of the PLZT light shutters. Sixty-four sample-hold circuits using J-FETs were connected to the amplified video output. PLZT light shutters were used as sample-hold capacitors in the circuit. The sample-hold generator, consisting of shift resistors, produced a sample-hold pulse, 70V $_{pp}$ high and 0.8μs wide. The pulse was supplied to the J-FET gates from No. 1 to No. 64 sequentially during the horizontal scanning period. All PLZT shutters were driven sequentially according to the pulse timing pattern and charged according to the video-signal voltage at the time when the pulse was applied.

In this way, PLZT shutters in the array were modulated with analog signals sampled from the video signal. The applied voltage on each PLZT shutter was held until the next horizontal scan to maintain light intensity. To reproduce a two-dimensional video image, the movable mirror was driven by a sawtooth wave synchronized to the vertical scanning.

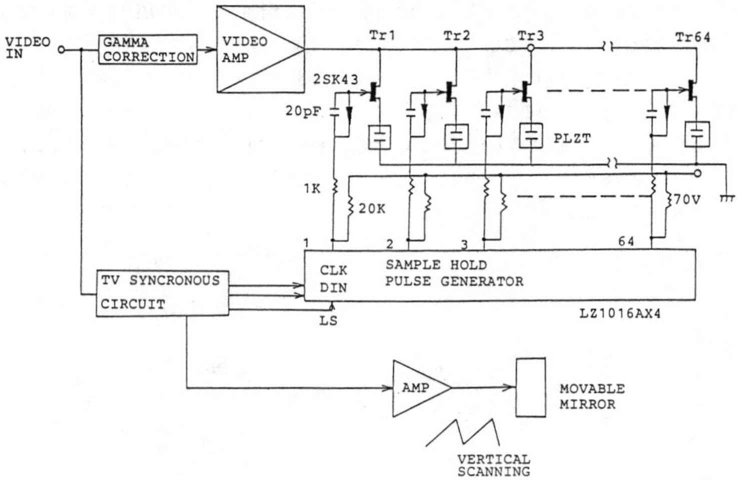

Fig. 10. PLZT Video Projector Circuit.

Figure 11 shows a test pattern projected onto the screen by the present projector. The possibility of a PLZT projector is apparent, but problems in brightness and the uniformity of image will have to be solved. The brightness on the screen was measured by an illumination photometer, and from 5 to 10 Lm was calculated as the luminuous flux on the screen. Since the entire luminous flux of the Xenon light source was about 10,000 Lm, the utility factor of the light was very low. The reasons for this are:

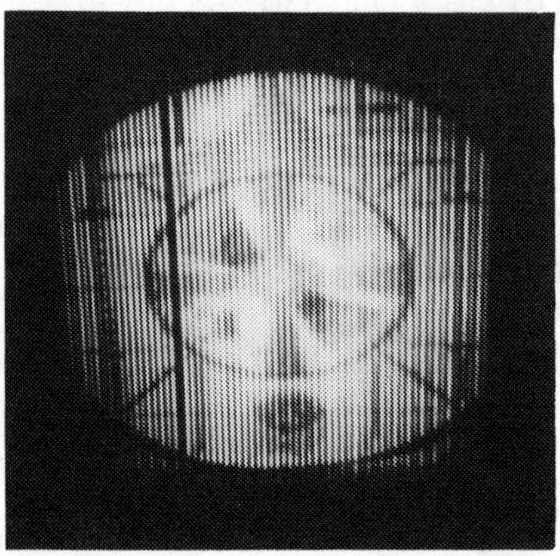

Fig. 11. Video Image Projected on the Screen.

① Only a small part of the light from Xenon can pass through the small diameter lens L1.
② The reflection loss of the PLZT array and lenses is high.
③ The absorption loss of the polarizer and analyser is high.
④ The light reduction system causes still more loss.
⑤ There is a loss caused by the formation of a slit shape light flux.

The projected image on the screen shown in Fig. 11 does not have a uniform brightness in the horizontal direction, and this is related to the half-wave voltage or contrast ratio of each shutter in the PLZT array. The half-wave voltage ($V_{\lambda/2}$) of the shutters was measured at 60 ℃ as X(average)=61.0V, 3σ=8.9V and $3\sigma/X$ =14.6% (n=62). If 61.0V is applied to all of the shutters, we can calculate that the variance in the light passing through shutters is less than 22.4% because the light is proportional to $\sin^2 V^2_{\lambda/2}$. It is important to control the variance of half-wave voltage in a PLZT shutter array to get a uniform brightness.

The holizontal resolution of the image in Fig. 11 is not sufficient for ordinary TV. This will be improved by using a high density linear shutter array which has more than 250 light shutters.

We had to abandon this development at work because there were too many problems to solve, especially the problem of brightness. To improve the utility factor of the light, a Xenon light source with a smaller plasma ball, a large size movable mirror with a quick response, and some special lenses must be developed.

DISCUSSION

There follows a discussion of the problems common to the application of PLZT light shutter to video displays.

(1) PLZT Composition

The driving frequency of the shutter is high because of the TV system. The shutters in the viewfinder and in the projector are driven at 5.25 kHz and at 15.75 kHz, respectively. Under a high frequency and high voltage drive, the PLZT temperature increases because a PLZT light shutter is a ceramic capacitor with a large tanδ loss. The PLZT operating temperature depends on the PLZT size, configuration, driving voltage and frequency. The viewfinder's operating temperature was around 90℃ the projector's was at 60℃ at room temperature.

A PLZT's electro-optic constant, which determines the driving voltage, decreases as the PLZT temperature increases, so at a higher temperature the shutters have to be driven at a higher voltage. On the other hand, a PLZT's electro-optic constant depends not only on the temperature but also the PLZT composition. 7.3/65/35 PLZT cannot be used as a shutter composition at room temperature because it has a memory effect at room temperature, however, it is the best shutter composition at around 90℃. To minimize the driving voltage a PLZT composition with the highest constant at the operating temperature should be chosen.

(2) High-voltage Drive and Power Consumption

Even if the optimum composition is chosen, the driving voltage of the shutter is more than 50 V. The shutters in the viewfinder and projector are driven at 170 V and 70 V. This high voltage drive prevents the application of PLZT light shutters to actual battery-drive electronic consumer products which use a battery of less than 6 V. High-voltage drive needs a special high-voltage supplier which makes the product bigger and the cost higher. Moreover, a high-voltage drive requires a higher power comsumption because the power comsumption is proportional to the square of the driving voltage. To extend PLZT shutter applications, some way should be found to make the driving voltage less than 10 V.

(3) Light Shutter Array

If the PLZT shutter is used as a display device, usually a light shutter array is desirable because of parallel light treatment. Both the viewfinder and projector use a light shutter array. To make a high-density light shutter array, it is important to solve the problems of crosstalk, assembly and wiring. We developed a new PLZT light shutter array using parallel-plate electrodes which has no crosstalk and is easy to make. We hope this type of PLZT array will be used not only for video displays but also other PLZT applications, for example, printer heads.[6)]

SUMMARY

This paper reports our two video displays, a color viewfinder and a linear light valve projector using PLZT light shutters, and discusses problems with the application of PLZT light shutters to video displays. In a video display, light shutters are driven at a high frequency because of the TV system.

A high-frequency and high-voltage drive causes the operating temperature of a PLZT to be high. We can solve this problem by choosing a PLZT composition which has the best electro-optic constant at the operating temperature.

It is difficult to solve the problem of the high-voltage drive which prevents the application of the PLZT to the battery-drive products. At present, the liquid crystal display is favored because it is driven at less than 5 V and is IC compatible. But a PLZT light shutter has many advantages compared to liquid crystal: a high-speed response less than 1 μs which is enough for video displays, a high contrast ratio of more than 1000, and reliability because of its solid-state creramics. In order to use a PLZT light shutter in a practical video display, we had to construct a stationary display with no high-voltage problem.

Recently a big video display like Sony's JumboTron* using CRs was made available. There are two types of big displays, one is this CRT type and the other a liquid-crystal type. The CRT type has a good response and a good contrast ratio but is expensive. The liquid-crystal type does not have a good response or contrast but is less expensive. A big display using PLZT light shutters may provide a good compromise.

* JumboTron is a trademark of Sony Corporation.

REFERENCES

1) M.Himuro, K.Murano, S.Nishigaki, T.Masatoki and K.Ohno, "A Color Viewfinder Using PLZT Light Shutters and a B/W CRT" JJAP, Vol 22 (1983) Supplement 22-2, pp.132-134

2) K.Murano, M.Himuro, Y.Watanabe, S.Matsumoto and S.Nishigaki, "A PLZT Light Shutter with Low Power Consumption" JJAP, Vol 22 (1983) Supplement 22-2, pp.129-131

3) K.Murano, M.Himuro, Y.Watanabe and M.Tonosaki, "A Video Projector Using a PLZT Light Shutter Array" JJAP, Vol 24 (1985) Supplement 24-2, pp.139-143

4) Cutchen, J.T., Harris, Jr.,J.O., and Laguna, G.R., "PLZT electrooptic shutters" Applications, Appl. Optics, Vol 14 (1975)pp.1866

5) Samek, N.E., and Munoz, M.A., "Light transmission characteristics of a high density electrooptical light gate array based on PLZT ceramics" Proc. SPIE. August 27-28,(1981) pp.307

6) Wessel, G. "Electro-optical ceramic (PLZT) as light-gate array for non-impact printer" Proceedings SPIE, Geneva, Switzerland (1983) pp.53-57

TWO-DIMENSIONAL DISPLAYS USING TRANSPARENT PLZT CERAMICS

Kenji UCHINO, Kosaku TOKIWA, Jayne GINIEWICZ
Department of Physics, Sophia University
Kioi-cho 7-1, Chiyoda-ku, Tokyo 102, Japan
Yasuhide MURAI and Kingo OHMURA
Devices Research Department, Laboratory, Fujitsu General Ltd.
Suenaga 1116, Takatsu-ku, Yokohama 213, Japan

ABSTRACT

A new type of two-dimensional light valve array for an image projector has been developed using transparent PLZT ceramics. A light shutter array with 10x10 pixels was fabricated by a sophisticated tape casting technique, starting from coprecipitated PLZT fine powder. Plate-through and separate internal electrodes are stacked alternately so as to make vertical addressing by an external electrode connecting separate electrodes and horizontal addressing by a plate-through electrode. A pair of shutter elements beside a plate-through electrode constitute an image pixel. The transmittance of 62% (at 633nm) could be obtained by atmosphere controlled sintering of a green chip, which is in good comparison with 63% for the ideal sample prepared by hot pressing. A contrast ratio of 80 is realized with a response time of 10μsec under a half-wavelength voltage of 100V applied. The applicative feasibility to a high definition image projector was verified.

INTRODUCTION

Recent requirements for high-definition TV's are remarkable, and several systems have been proposed. There are three typical types of the TV as shown in Fig. 1; 1)illumination type, 2)reflection type and 3)projection type. A CRT TV exemplifies the first category, whose large volume (especially wide depth) unfortunately shares too large space of a Japanese "small" room, in addition to its costly price. Reflection types include a liquid crystal TV, the demerit of which is its dark image. On the contrary, a projection type has several advantages such as changeable image size and rather low price.

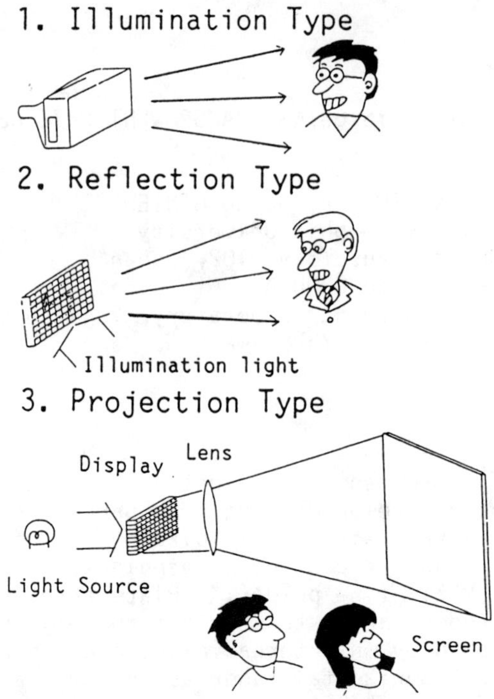

Fig. 1 Three typical types of TV; 1)illumination type, 2)reflection type and 3)projection type.

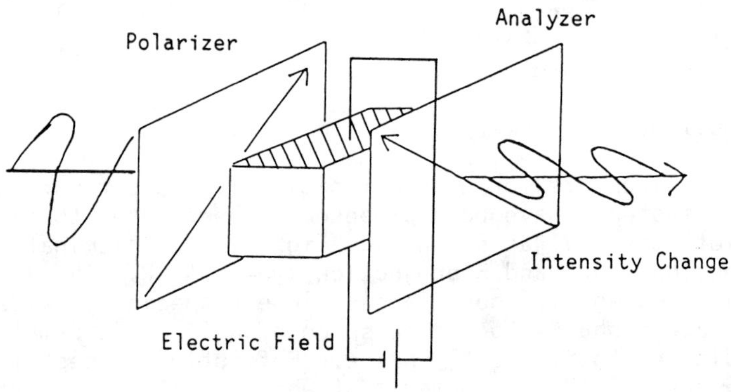

Fig. 2 Fundamental construction of an electrooptic light shutter.

This paper deals a concept of a projection type TV utilizing two-dimensional PLZT displays. Electrooptic characteristics in PLZT ceramics are generally much superior in response speed and contrast ratio to those in liquid crystals. Moreover, durability of ceramics to a strong light illumination is excellent. However, the most significant problems in PLZT devices, which have prohibited their actual commercialization, are the high driving voltage and the cost. Therefore, the development of a simple mass-production process and the designing of electrode configurations with a narrow gap will be the key factors for PLZT displays. In this paper we will describe a possible solution to overcome most of the above-mentioned problems.

PLZT ELECTROOPTIC DEVICES

Lead lanthanum zirconate titanate (abbreviated as PLZT) ceramics are highly transparent and exhibit a strong electrooptic response. Especially La-rich compositions such as 9/65/35 (La/Zr/Ti ratio) exhibit the second-order electrooptic effect (Kerr effect); that is, the birefringence Δn is induced only when an electric field E is applied.

$$\Delta n=-(1/2)Rn^3E^2 \tag{1}$$

where R is the quadratic electrooptic coefficient and n is the original refractive index of the PLZT ceramics.

When the sample is placed between crossed polarizers (Fig. 2), the transmitted light intensity is represented by the following equation:

$$I=I_0A\ \sin^2\{(Rn^3t/2\lambda)E^2\} \tag{2}$$

where I_0 is the incident light intensity, A is an equipment constant, t is the thickness of the PLZT, and λ is the wavelength of the light.

Typical conventional constructions of one-dimensional and two-dimensional light valve arrays are illustrated in Fig. 3. The 1-D device in Fig. 3(a) is produced by a highly laborious process involving cutting, electroding, assembling and wiring of a sintered ceramic, and exhibits a rather dark image because the line image is scanned on the screen[1]. The 2-D device in Fig. 3(b) shows slow response due to high capacitance of a long unit cell and asterismal light leak like the "Southern Cross" is also a significant problem (the authors' unreported work).

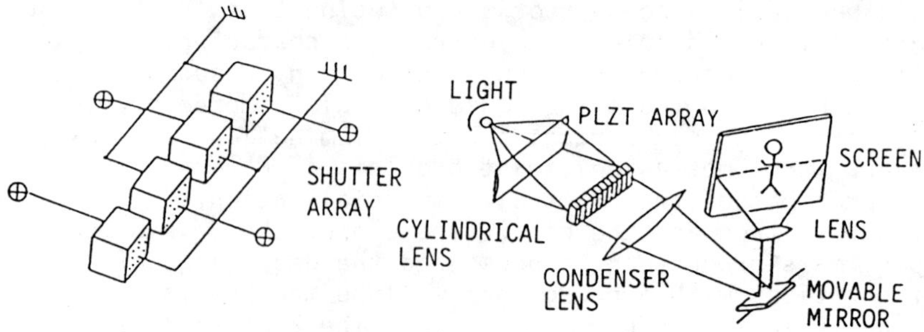

(a)

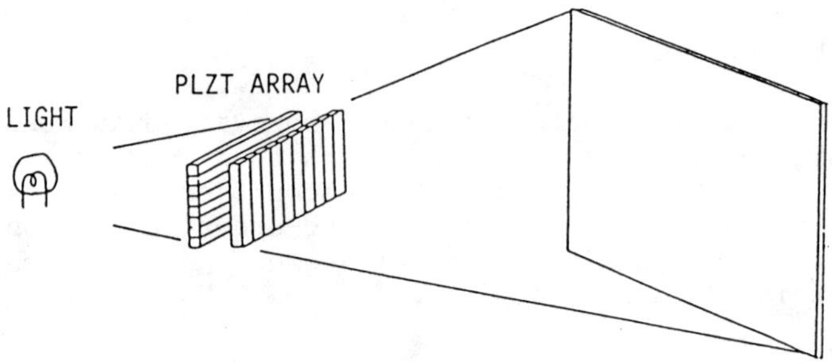

(b)

Fig. 3 Conventional construction of (a)1-dimensional and (b)2-dimensional light valve arrays.

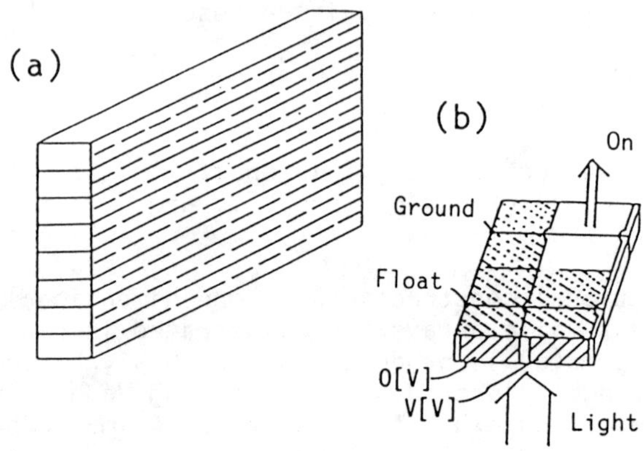

Fig. 4 Newly developed design of a 2-dimensional display.

On the contrary, newly developed design as shown in Fig. 4 presents a very bright image with no crosstalk-related problems and is easy to produce.

FABRICATION PROCESS OF THE 2-D PLZT LIGHT VALVE

The fabrication process of the two-dimensional PLZT light valve array is outlined in Fig. 5. Wet-chemically prepared (coprecipitated) PLZT 9/65/35 ($Pb_{0.91}La_{0.09}(Zr_{0.65}Ti_{0.35})_{0.98}O_3$) powders were mixed with organic solvent and binder and formed into a green sheet. Platinum internal electrodes were printed on the green sheets. The electroded sheets were then laminated alternately in 90° different orientations under a pressure of 3000 psi, so as to make plate-through and separate electrodes. The laminated body was sintered in an oxygen-controlled atmosphere, and the bulk was cut and polished. Finally the external connecting electrodes were applied to make vertical and horizontal addressing.

Figure 6 shows the SEM picture of the PLZT ceramic, where the average grain size is about 9μm. The grains should be grown more than 3μm in diameter, because the electrooptic response is radically decreased below 3μm (see Fig. 7)[2].

Figure 8(a) shows the electrode configuration of a (10x10) matrix PLZT light valve. The shaded portion of the device in the figure represents one image unit (pixel). The vertical separate internal electrodes were connected by external electrodes printed on the surface of the device. The horizontal plate-through electrodes were embedded 100μm deeper from the optical surface to avoid shorting with the vertical electrode connectors. Figures 8(b), 8(c) and 8(d) shows pictures of the newly fabricated display. Note that the layer thickness is about 0.35um.

CHARACTERISTICS OF THE PLZT LIGHT VALVE

The optical transmittance of the PLZT device fabricated by the tape casting technique is plotted as a function of light wavelength in Fig. 9. The transmittance of 62% could be obtained by atmosphere-controlled sintering, in good comparison with 63% for the ideal sample prepared by hot-pressing.

The brightness for red, green and blue light was measured as a function of the applied voltage (Fig. 10), where the electrode gap was 0.45mm. The contrast ratio defined by a ratio of brightness on the screen under half-wavelength voltage applied over brightness under zero volt applied (220(cd/m^2)/2.8(cd/m^2)),

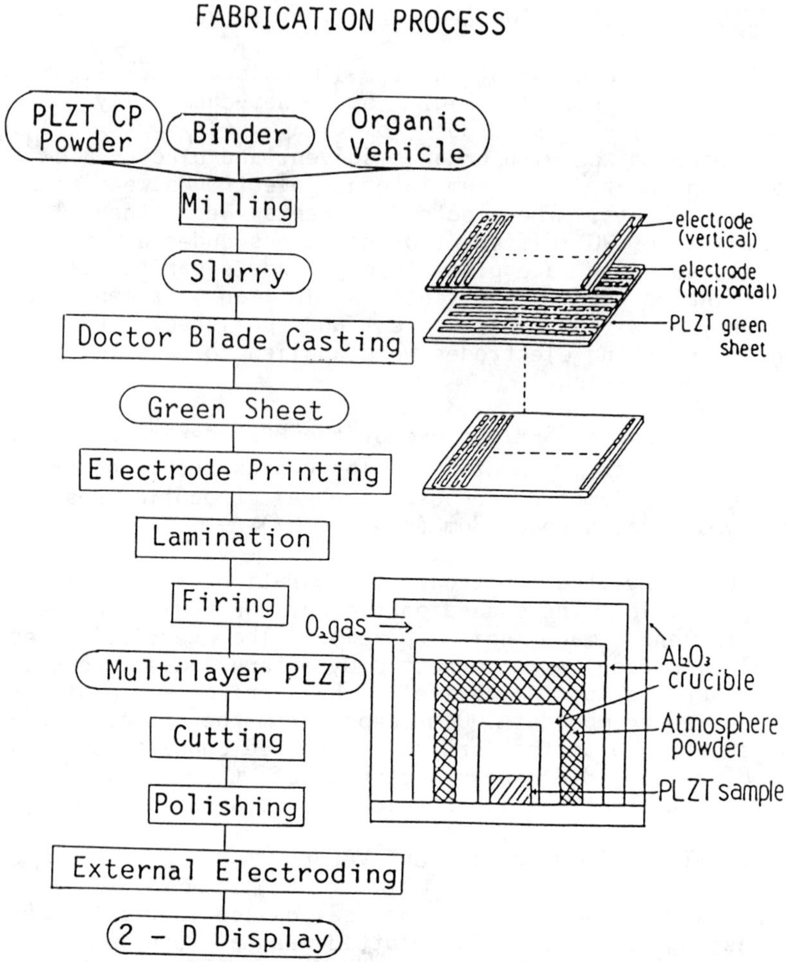

Fig. 5 Fabrication process of the newly developed 2-D display.

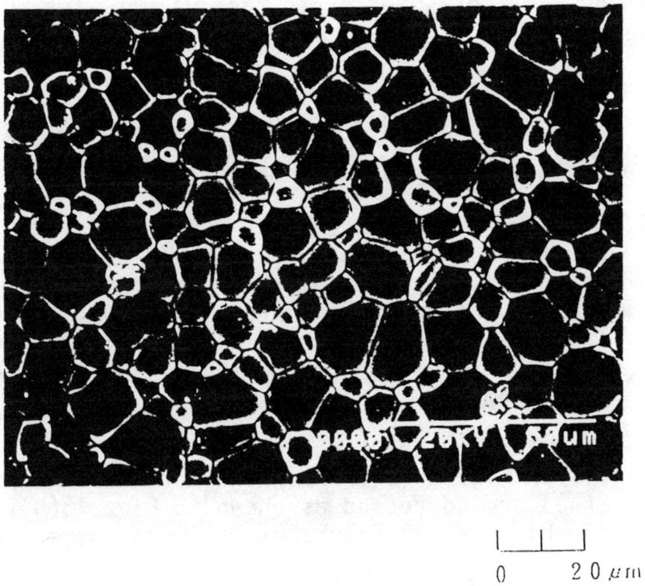

Fig. 6 SEM picture of the PLZT surface sintered in an oxygen-controlled atmosphere.

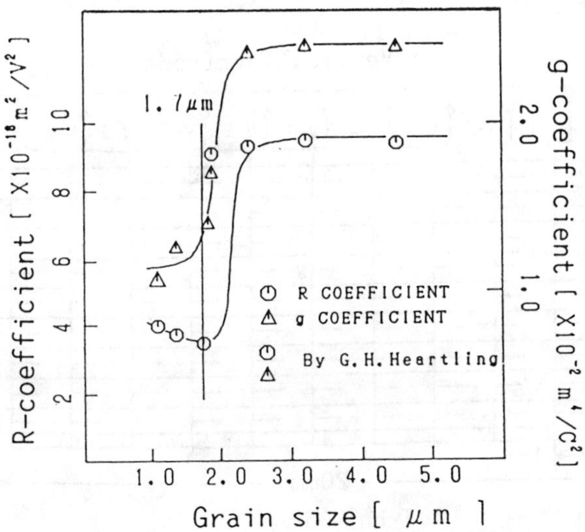

Fig. 7 Grain size dependence of the electrooptic coefficients, R and g, in PLZT 9/65/35.

is about 80, rather superior to the values for conventional displays.

The response associated with a single pixel of the display was monitored for a stepwise voltage (Fig. 11). The response time is less than 10μsec, which is rapid enough to drive the PLZT shutter at a raster frequency of conventional cathode ray tubes.

CONSTRUCTION OF THE COLOR PROJECTOR

The construction of the PLZT color image projector is shown in Fig.12. Three light fluxes, Red, Green and Blue, are obtained by separating Xenon light with dichroic mirrors, and passed through each identical PLZT shutter. Then, the fluxes are superimposed to make a color image.

The driving circuit of the PLZT display is represented schematically in Fig. 13(a). When the terminals of the device (see Fig. 8(a)) are addressed as shown in Fig. 13(b), the image appearing in Fig. 13(c) (alphabet "F") is generated on the screen.

The performance characteristics of the newly developed PLZT image projector are summarized in Table 1.

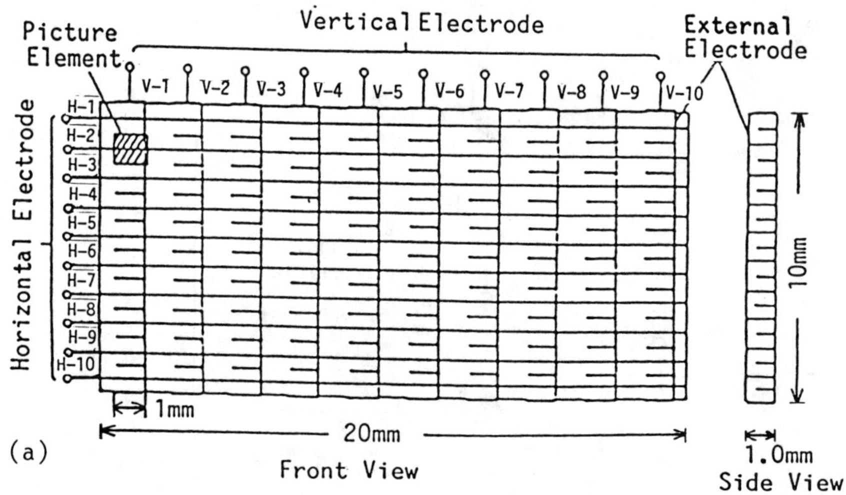

Fig. 8 (a)Schematic electrode configuration of a (10x10) matrix PLZT light valve.

(b) 

(c) 

(d) 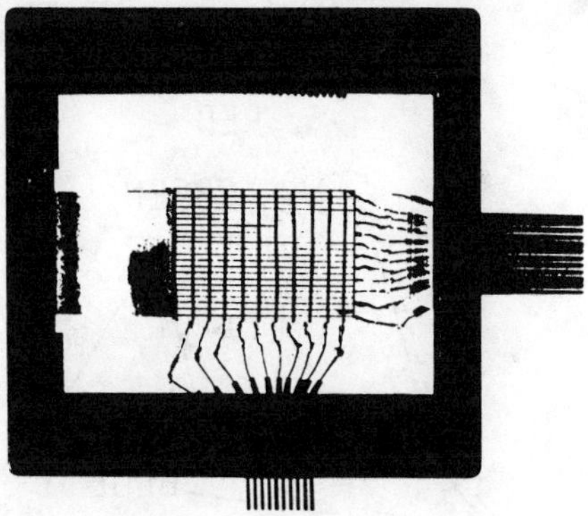

Fig. 8 (b)Top view of the actual electrode configuration of a light valve. (c)Tilted view. (d)A PLZT light valve with external electrodes.

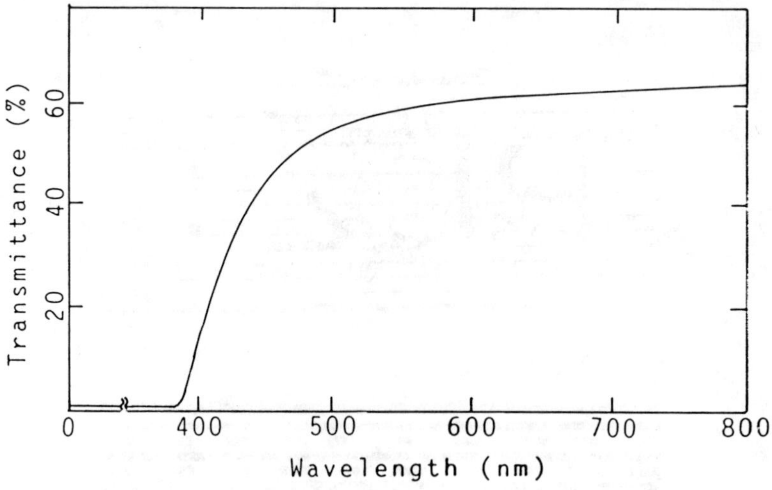

Fig. 9 Optical transmittance of the PLZT device fabricated by the tape casting technique.

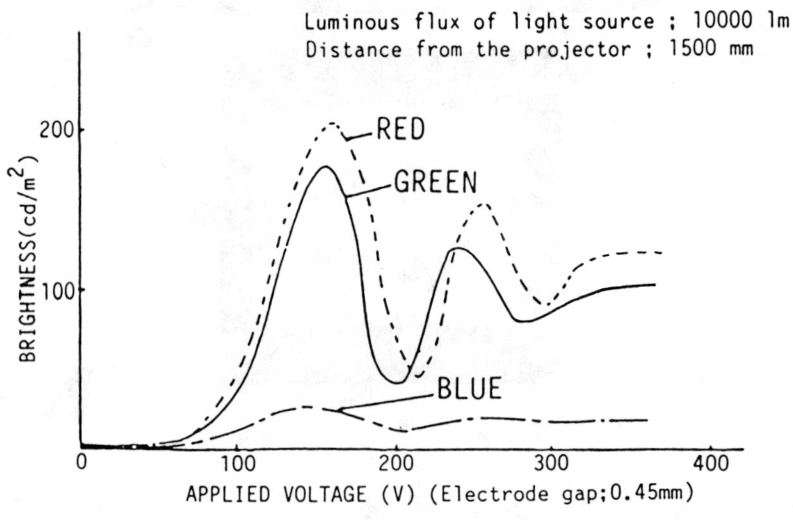

Fig. 10 Brightness on the screen versus applied voltage for red, green and blue light (electrode gap : 0.45mm).

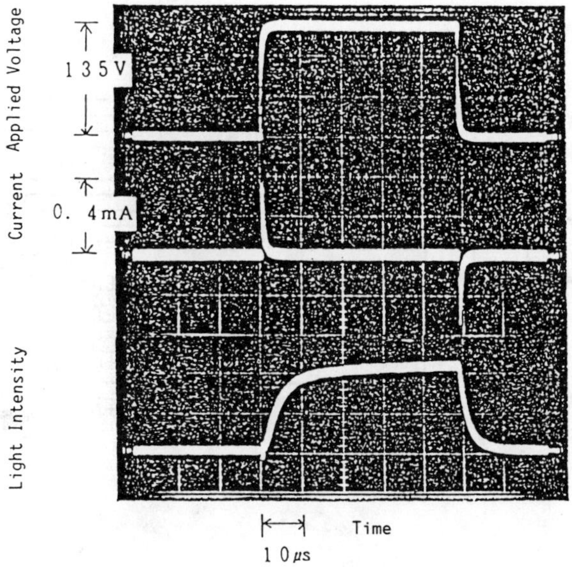

Fig. 11 Light valve response for a stepwise voltage.

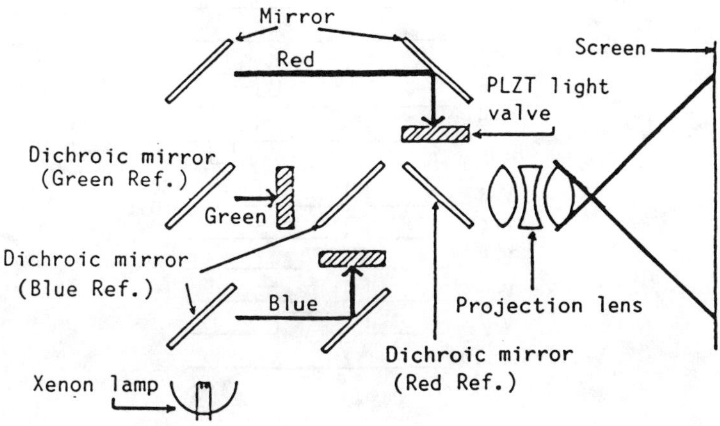

Fig. 12 Construction of the PLZT color image projector.

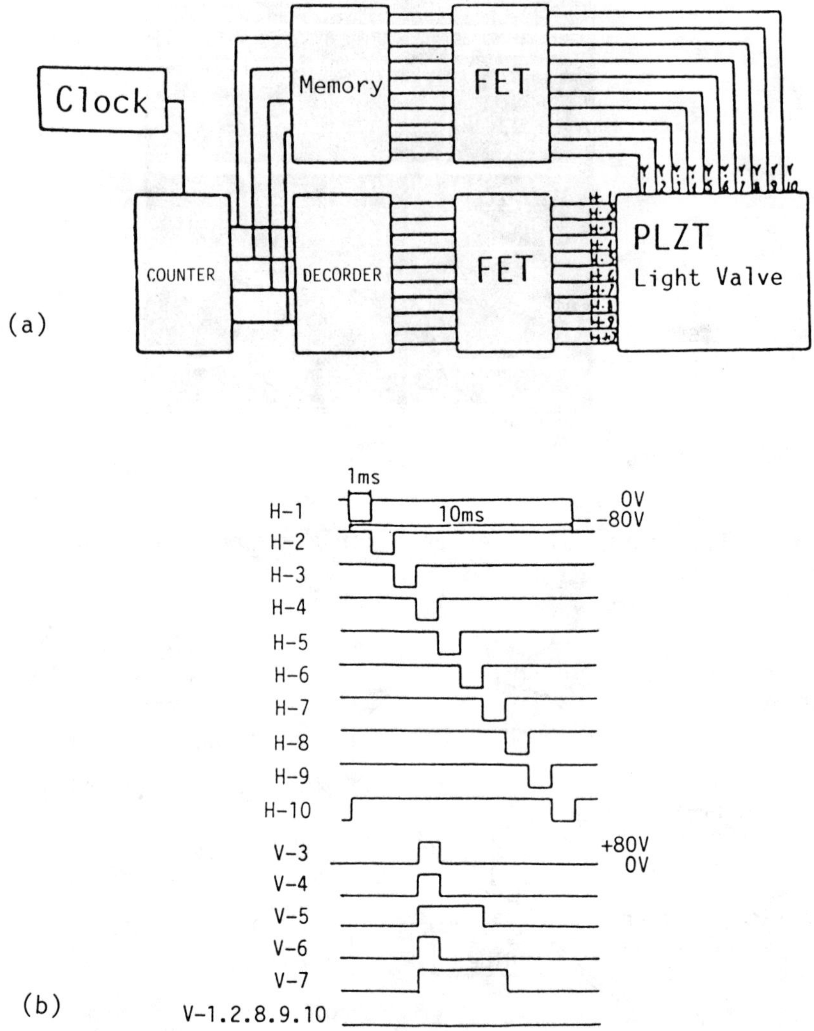

Fig. 13 (a)Driving circuit of the PLZT display. (b)Example waveforms of the driving signals.

(c) 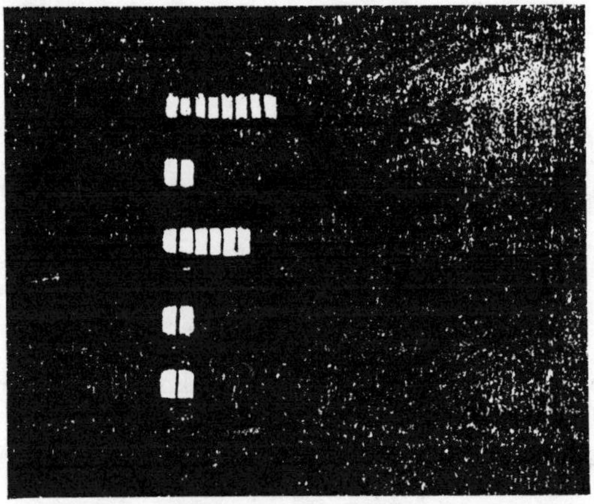

Fig. 13 (c)An example image on the screen from the PLZT projector.

Table 1 Performance characteristics of the new PLZT projector.

- Pixel
  - Size : 1.40x1.40mm
  - Optical Path (Thickness) : 1.0mm
  - Electrode Gap : 0.35mm
  - Pixel Pitch : 2.0mm
- No. of Pixels : 10x10
- Contrast Ratio : 80
- Response Time : 10μsec
- Display Viewing Area : 20inch
- Total Luminous Flux (Xenon) : 10^4 lm
- Brightness : 80 cd/m^2 (nt)

CONCLUSIONS

A new type of PLZT 2-dimensional light valve fabricated by a tape casting technique has been developed and a prototype color image projector incorporating this light valve was investigated.
This light valve exhibits bright image with no crosstalk-related problems. It can be driven in quick response(10μsec) by a rather low drive voltage (100V/0.35mm gap) in comparison with conventional PLZT devices. Moreover, the new device is easy to mass-produce, leading to a low manufacturing cost.

It is worth to note the technology backgrounds for success of the new 2-D display. Firstly, easy production process for transparent PLZT ceramics has been developed in these few years, starting from sophisticated ultrafine raw powders without using a hot-press furnace. Secondly, we have improved tape casting technology for producing large 2-D ceramic displays without any shape distortion generated during sintering process.

Further research will be required for fabricating a high definition image projector.

REFERENCES

1. K. Murano, M. Himuro and M. Tonosaki: Jpn. J. Appl. Phys., 24, 139 (1985)
2. K. Tokiwa and K. Uchino: Ferroelectrics, 94, 87 (1989)

INDEX